Organic Electronics 1

Series Editor
Robert Baptist

Organic Electronics 1

Materials and Physical Processes

Thien-Phap Nguyen

iSTE

WILEY

First published 2021 in Great Britain and the United States by ISTE Ltd and John Wiley & Sons, Inc.

ISTE Ltd
27-37 St George's Road
London SW19 4EU
UK

www.iste.co.uk

John Wiley & Sons, Inc.
111 River Street
Hoboken, NJ 07030
USA

www.wiley.com

Library of Congress Control Number: 2020948483

British Library Cataloguing-in-Publication Data
A CIP record for this book is available from the British Library
ISBN 978-1-78630-321-9

Contents

Introduction

Organic electronics can be defined as a branch of general electronics that focuses on studying the properties and applications of organic semiconductors. These materials, referred to in common usage as plastics, are, in fact, a distinct class separate from that of ordinary plastic materials. They may be small molecules, or conjugated polymers, which have an electronic structure comparable to that of conventional semiconductors. Therefore, their physical properties are very similar to the properties of these latter materials. However, organic materials are generally amorphous and their electrical conductivity is much lower than that of traditional semiconductors. As a result, they have been neglected for a long time despite the discovery of some of their notable physical properties, such as the discovery of electroluminescence in anthracene in the 1960s.

In all scientific or technological disciplines, an idea, concept or creation can arise with a change or progression and profoundly alter the course of the evolution of the discipline – and sometimes even the course of the history of science. Such was the case in 1947 when the researchers John Bardeen, William Shockley and Walter Brattain invented the transistor, a device that would revolutionize traditional electronics. This invention is what first allowed for the design of circuits that provided logical functions, then for these circuits to be integrated into complex systems capable of handling sophisticated tasks, and finally for these miniaturized systems to be incorporated into the portable electronic devices that we now use every day. A similar progression occurred in organic electronics in 1987, when Tang and Van Slyke used the evaporation of small molecules in a vacuum to create diodes that emitted light in the form of a thin film that operated with a turn-on voltage lower than 10 V. Their work demonstrated that, in practice, organic devices are suitable for optoelectronic applications in the same way as their inorganic counterparts. A few years later, Burroughs *et al.* (1994) demonstrated that by using conjugated polymers deposited as thin films from a solution, they could also produce light-emitting diodes with a low operating voltage. This work paved the way for film depositing techniques using solutions and led to the production of devices through printing developed later in laboratories. For this domain, we will use the descriptors "flexible" or "printed" to indicate the discipline of organic electronics. Advances in research have been very rapid and have shown remarkable results, not only in the understanding of physical processes in materials and components but also for the creation of new electronic devices now available on the market. Today, organic electronics has become its own field, one that will prove to be important technologically and economically in the near future.

In France, education on organic electronics at the university level is developing, but remains limited in comparison with other European countries. It should be noted that these courses are generally offered in universities that carry out industrial research and development activities on organic electronic materials and devices. With regard to the books written on this discipline, there are very few titles available in French.

This book, divided into two volumes, was written with the goal of introducing organic electronics to students and researchers who are interested in this new discipline. It is organized into the following sections: Chapter 1 of Volume 1 provides an overview of the basic notions of the theory of traditional semiconductors, of which some of the material presented will be used later on. The materials used for the creation of devices are described in Chapter 2. The physical processes, which take place in the volume and interface of the layers of devices, are presented and explained in Chapters 3, 4 and 5. Volume 2 presents the primary applications of organic materials in optoelectronic devices. The first chapter of this volume centers on organic light-emitting diodes (OLEDs), the second on organic solar cells (OSCs or OPVs) and the third on organic field-effect transistors (OFETs). Chapter 4 deals with the practical and economic aspects of the industrialization of organic components. It also includes a discussion on the environmental aspects of the use of organic materials and devices.

As the title indicates, this book is not intended to provide a detailed and complete description of the materials, physical processes and applications involved in organic electronics. This would be far beyond the scope of what can be addressed within a single book. The choice of topics and of the extent to which the subjects are discussed were made by the author, based on his own experience from research and as an instructor. Interested readers will be able to find more detailed discussions on certain topics using the references provided. A list of acronyms used in the text is also included at the end of this book.

I would like to thank the people who devoted their time to proofreading and providing comments and suggestions, which allowed me to improve the writing of this book and the way in which certain areas are presented: Philippe Lerendu, Serge Lefrant, René Leparoux, Agnès Bournigal-Giret, Maxime Bayle and Jean-Luc Duvail. I would also like to thank ISTE for proposing this project, which I hope will help to arouse public interest in the new and promising possibilities offered by organic electronics in the scientific and technological communities in France and around the world.

1

Semiconductor Theory

This chapter provides a brief overview of the basic concepts of semiconductor theory with regard to particular physical characteristics that allowed the creation of electronic components that would revolutionize various technologies starting in the middle of the 20th Century. We will now describe the operation of P–N junctions, which are the essential components for the creation of electronic devices. This basic knowledge will then allow us to address physical processes in organic materials and then to understand the operation and use of devices based on new organic semiconductors.

1.1. Introduction

In electronics, the basic materials are semiconductors (SCs). They differ from metals in their dependence on the temperature of their electric characteristics. Essentially, when the temperature increases, the resistivity of metals increases, while the resistivity of SCs decreases. The theoretical contributions made by quantum mechanics allowed the electrical properties of SCs to be studied and explained based on the theory of energy bands. Unlike metals, in which the charge carriers are electrons, SCs can also carry positive charges, known as *holes*, that contribute to the electrical conductivity of these materials. The concentration of the carriers can also be changed by "doping", that is, the incorporation of the selected impurities in a determined quantity. This process not only allows the conductivity of the SC to be changed, but also favors the transport of one type of charge carrier at the expense of the other. An SC is characterized as N-type if the majority of the carriers are electrons and as P-type if the majority of the carriers are holes.

The combination of an N-type and P-type SC is a P–N junction, which is the fundamental element of all traditional electronic components. The assembly of these elements, prepared with other materials, makes it possible to create components that perform special functions, and the design of the circuits with these components leads to multiple applications of various different types.

The miniaturizing of components has given rise to the field known as *microelectronics*, the technology for microscale component manufacturing while maintaining the same characteristics as conventional components. This new technology has significantly reduced the size of miniaturized electronic circuits, as well as certain devices that can be made portable. With the advent of integrated circuits, discrete components can now be installed on boards using a minimum amount of space and electrical connections. This evolution in the size of electronic components follows the projections of Moore's law, which holds that the density of the components able to be installed will double about every two years. At the same time, this reduction in size leads to a reduction in the amount of material used. With the introduction of new digital technologies, and especially nanotechnology, technological advances in electronics seem to be limitless. However, this progress occurs more slowly as the size of the components decreases and, once they shrink to the size of a few tens of nanometers, the components can no longer be guaranteed in terms of their size and quality. It then becomes a question of modifying or changing the material used in the SC or the structure of the components, or better yet, the way they are built.

In the late 1980s, a scientific paper by an American team reported the results of measurements made on a light-emitting diode using, instead of a crystalline SC, an organic material: tris (8-hydroxyquinoline) aluminum (III) or Alq_3. The article gave the first-ever description of the use of a non-crystalline SC in a device that was both practical and functional, as the diode was able to emit light under an applied voltage of about 10 volts. What was unique about this experiment was its use of a thin film structure for the active material, a first in the design of electronic components and devices. A few years later, researchers in the UK published their discovery of electroluminescence in *conjugated polymers*, which spurred on a considerable number of studies on organic SCs, both with regard to the study of materials and the study of the particular physical processes of the devices based on organic SCs. The enthusiasm of the scientific community for this new field of electronics allowed for significant advances in knowledge and applications to be made over a period of about ten years. Display screens entered the market in the early 2000s, followed shortly thereafter by organic photovoltaic panels. This interest arose due to the promising possibilities offered by organic electronic devices, which first appeared at a time when the problems in traditional electronics were beginning to become clear.

In this book, we will present the fundamental notions of organic electronics along with those of electronics based on crystalline SCs, as well as certain notable differences that correspond to the specific properties of organic materials. We will also focus on the most recent applications of this new field of electronics, which do not compete with traditional electronics, but instead complement and reinforce them to a certain extent.

We will present several basic concepts from the study of crystalline SCs (CSCs) in order to address the various different aspects of organic SCs (OSCs). It should be note that, for reasons of clarity, these abbreviations will be used to distinguish the two materials when there may be a possible confusion in the description. Otherwise, we will use the abbreviation SC to refer to the SC component. We will review the essential and useful points on the properties of CSCs, as well as P–N and metal/SC junctions. The principle of electronic devices based on OSCs or CSCs will be discussed in Chapters 1, 2 and 3 of Volume 2.

1.2. Review of the basic concepts of crystalline semiconductors

SCs used in traditional electronic devices are crystals, that is, materials made from atoms, ions or molecules arranged in an orderly and repetitive manner in a three-dimensional network. As a result, the charge carriers in motion within these materials are subjected to a periodic potential field, and their energy is restricted to energy domains, called *allowed bands*. This concept introduces the notions of the conduction band (CB), the valence band (VB) and the forbidden band or energy gap. The arrangement of allowed bands and the gap value allow us to distinguish conductive materials or metals, insulators and SCs.

In a conductor, the density or concentration of electrons is of the same order of magnitude as the density of atoms, while in an insulator, the ratio of the densities of free electrons to atoms is less than 10^{-20}. An SC can be considered as a bad conductor at room temperature if its electrical conductivity varies between 10^{-10} and 10^3 S/cm, whereas the conductivity of metals varies between 10^4 and 10^6 S/cm. This poor conductivity results from the low density of electrons in the CB and the holes in the VB, because the energy required for creating an electron–hole pair (or exciton) using thermal generation is relatively large for most SCs. To explain the physical properties and processes of SCs, we will use silicon (Si) as an example, noting that its properties and processes are applicable to other SCs.

1.2.1. *Intrinsic semiconductors*

In a lattice of pure silicon, the bonds between atoms are covalent. Each Si atom is surrounded by four neighboring atoms that exchange valence electrons with it, so that all atoms have eight valence electrons. This configuration ensures the stability of the silicon crystal lattice. A number of electrons that have acquired sufficient energy (provided by the energy supply from an external source that can be heat, electricity or light) can break their bonds and become free. An electron released in this way passes from the VB to the CB and creates a hole in the VB through the *generation process*.

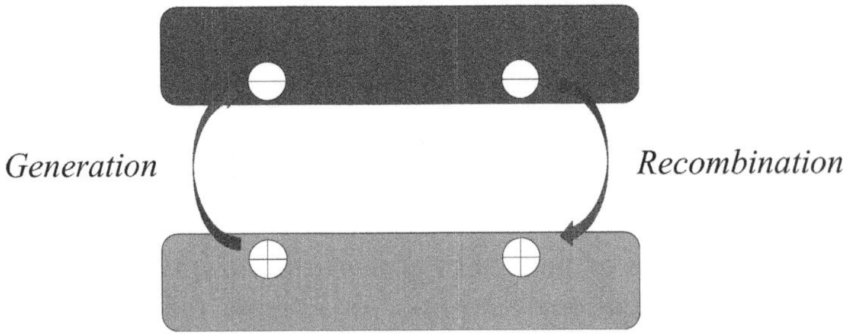

Figure 1.1. *Generation and recombination process*

Let n_i and p_i be the concentrations of electrons and holes of the intrinsic SCs. We can write $n_i = p_i$ because there are as many electrons in the CB as there are holes in the VB that are created through the generation process. The intrinsic concentrations verify the *mass action law* as follows:

$$n_i \times p_i = n_i^2 \quad [1.1]$$

The electrons and holes thus created in the SC can then disappear by *recombination*. This process allows an electron in the CB to take the place of a hole in the VB by losing an amount of energy theoretically equal to the energy of the SC gap. This energy can be recovered either thermally (phonon emission) or optically (photon emission). When the SC is in thermal equilibrium, the concentrations of CB electrons and VB holes are stable because the rates or speeds of generation and recombination of electron–hole pairs are constant. This concentration is a function of temperature *T*, governed by the relation:

$$n_i = p_i = A\,T^{3/2}\,exp\left(-\frac{E_G}{2\,k\,T}\right) \qquad [1.2]$$

where A is a constant, E_G is the SC gap and k is Boltzmann's constant.

1.2.2. *Extrinsic semiconductors*

By adding impurities to the intrinsic SC, the concentrations of electrons and holes, and therefore their electrical conductivity, can be controlled. In comparison with metals, SCs contain fewer charge carriers, but the fact that their concentrations can be adjusted to desired values is technologically essential for electronic device applications.

For silicon, a "doping" using the elements from group V of the periodic table helps to promote conductivity by electrons, while reducing the number of holes. Each atom of impurity, called a *donor* (*D*), offers the possibility of providing a free electron to the silicon CB when it is ionized. The doped SC is referred to as *N-type*. Let N_D be the concentration of donors and N_D^+ the concentration of ionized donors. The concentration of free electrons due to doping in the CB of the SC is equal to $n = N_D^+$. Similarly, the use of an element from group III of the periodic table promotes conductivity through the holes. After ionization, a doping atom, known as an *acceptor* (*A*), provides a hole for the VB and the SC is called *P-type*. Let N_A be the concentration of acceptors and N_A^- the concentration of ionized acceptors. The concentration of free holes due to doping in the VB of the doped SC is equal to $p = N_A^-$.

The ionization of impurities is a thermally activated process, and in principle, the concentration of charge carriers in the bands depends on the temperature of the SC. However, the ionization energy of typical impurities is low (of the order of several tens of milli-electron volts, or meV) so that they are virtually all ionized at room temperature (T=300 K). At this temperature, the concentration of electrons due to doping of an N-type SC is therefore $n = N_D$ and the concentration of holes due to doping of a P-type SC is therefore $p = N_A$.

In a doped SC, the creation of electron–hole pairs by generation and recombination processes occurs at the same time as the ionization of impurities. The total concentration of electrons in the CB is the sum of two concentrations: one due to the ionization of donors N_D^+ and the other due to generation/recombination that we denote as n_0. Similarly, the total concentration of holes in the VB is the sum of two concentrations, N_A^- and p_0.

The charge neutrality condition is written as:

$$N_A^- + n_0 = N_D^+ + p_0 \quad [1.3]$$

At room temperature, the ionization of impurities is practically achieved, and the concentrations of electrons and holes due to doping are $N_D^+ \sim N_D$ and $N_A^- \sim N_A$. At thermal equilibrium, the product of the concentrations of electrons and holes confirms the mass action law [1.1], but since the concentration of carriers created by ionization is dominant, the concentrations n_0 and p_0 are no longer equal. For an N-type SC, equations [1.1] and [1.3] result in:

$$n_0 \times p_0 = n_i^2 \quad [1.4]$$

$$N_A + n_0 = N_D + p_0$$

As $N_D \gg n_i$ and $N_D \gg N_A$, we therefore obtain:

$$n_0 \sim N_D$$

$$p_0 \sim \frac{n_i^2}{N_D} \quad [1.5]$$

For a P-type SC, we obtain:

$$p_0 \sim N_A$$

$$n_0 \sim \frac{n_i^2}{N_A} \quad [1.6]$$

1.2.3. *Fermi level*

The Fermi level (E_F) is defined as the highest energy level occupied by electrons at temperature $T_0 = 0\,K$. It allows us to determine how the electrons in an SC are distributed at a given temperature in the band diagram. For instance, at room temperature, the Fermi level of an intrinsic SC is located approximately in the middle of the energy gap. For an N-type SC, it is close to the bottom level E_C of the CB, and for a P-type SC, it is close to the top level E_V of the VB.

1.2.4. *Charge transport in semiconductors*

In an SC, charge carriers are transported by two mechanisms that allow the flow of the electric current. The currents corresponding to these mechanisms are the drift current and the diffusion current.

1.2.4.1. *Drift or electric current*

When no electric field is applied, free charge carriers move through the SC through thermal agitation and collide with other carriers. The average speed of all carriers is zero and no current flows in the SC. When an electric field $\vec{E}$ is applied, a carrier of charge q and mass m^* is subjected to an electric force $\vec{F} = q \times \vec{E}$. By taking $t = 0$ at the moment of a collision and applying Newton's second law of motion to the charge carrier, the equation for its instantaneous velocity is obtained:

$$v(t) = \frac{q \times E}{m^*} t$$

We can then deduce the expression of the average velocity or *drift velocity* of the free carriers:

$$\langle v \rangle = \frac{q \times E}{m^*} \tau_C \qquad [1.7]$$

where τ_C is the time between two collisions or the lifetime of the carriers.

Expression [1.7] indicates that the velocity of the carriers is proportional to the applied electric field. It can also be written as:

$$\langle v \rangle = v = \mu E \qquad [1.8]$$

where μ is the "mobility" of the charge carrier. It expresses the ease of movement of a charge carrier in the material and equals its velocity under a unit electric field. The current density J or the amount of charges that flows per second through a unit area of the SC with a concentration of carriers ρ is:

$$J = q\,\rho\,v = q\,\rho\,\mu\,E \qquad [1.9]$$

Expressions [1.7], [1.8] and [1.9] apply to both types of carriers. For electrons of a concentration n, the current density is:

$$J_n = -\,|e|\,n\,v_n = -\,|e|\,n \times (-\mu_n\,E) = |e|\,n\,\mu_n\,E$$

and for holes of a concentration p, the current density is:

$$J_p = |e|\,p\,v_p = |e|\,p\,\mu_p E$$

The total density of the drift current is:

$$J_C = |e|\,(n\,\mu_n + p\,\mu_p)E \qquad [1.10]$$

By introducing electrical conductivity using Ohm's law, we can write:

$$J_C = \sigma\,E$$

Thus, we obtain:

$$\sigma = |e|\,(n\,\mu_n + p\,\mu_p) \qquad [1.11]$$

1.2.4.2. *Diffusion current*

When charge carriers are not uniformly distributed in the SC, the carriers flow from high concentration regions to low concentration regions.

The diffusion of the carriers tends to lead to a uniform spatial concentration of charges in the material. The resulting current is called the *diffusion current* and the density of this current is governed by *Fick's law.*

For electrons moving along the *Ox* axis:

$$J_{Dn} = -\,q\,D_n \frac{\delta n}{\delta x} = |e|\,D_n \frac{\delta n}{\delta x} \qquad [1.12]$$

For holes moving along the Ox axis:

$$J_{Dp} = -q\,D_n \frac{\delta p}{\delta x} = |e|\,D_n \frac{\delta p}{\delta x} \qquad [1.13]$$

where D_n and D_p are, respectively, the diffusion coefficients of electrons and holes. These coefficients are related to the mobility of carriers through the Einstein relation, which is written as:

$$\frac{D_n}{\mu_n} = \frac{D_p}{\mu_p} = \frac{kT}{|e|} \qquad [1.14]$$

1.3. P–N junction

A P–N junction is obtained by allowing an N-type SC with a donor concentration of N_D to contact a P-type SC with an acceptor concentration of N_A. Since the concentrations of electrons and holes are very different in both types of SC, the majority charge carriers migrate from one SC to another by means of diffusion.

At thermal equilibrium, an internal electric field $\vec{E_i}$ is established at the junction and disrupts the diffusion. A space charge region (SCR) of a width λ, limited by the abscissa $(-x_p)$ and (x_n), is created at the junction in which no free charge is present.

1.3.1. *Space charge region*

From the point of view of energy bands, since the Fermi energy level E_{Fp} of the P-type SC is lower than the Fermi level E_{Fn} of the N-type SC, the electrons of the N-type SC will move from the N region to the P region until the Fermi level is aligned; thus, the probabilities of energy level occupancy by the electrons are identical in the two SCs.

The same applies for holes. Therefore, the Fermi levels E_{Fn} and E_{Fp} are aligned when an equilibrium is reached in the junction.

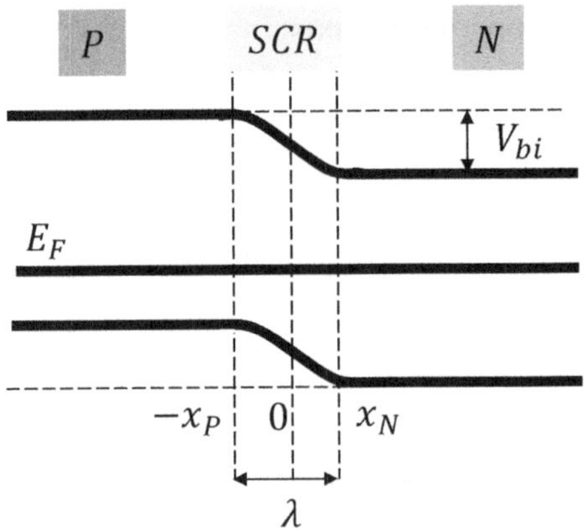

Figure 1.2. *Energy band diagram of a P–N junction. For a color version of this figure, see www.iste.co.uk/nguyen/electronics1.zip*

The internal electric field $E_i(x)$ can be determined by Gauss's law and the potential $V(x)$ in the junction by the relation:

$$E_i(x) = -\frac{dV(x)}{dx} \qquad [1.15]$$

When the junction is at thermodynamic equilibrium, the total current density $J_T = J_C + J_D$ is zero within the SCR. Using the Einstein relation [1.14] for the electrons and holes, and the expression of the law of mass action [1.4], it can be shown that the difference in energy between the two regions P and N is equal to:

$$\Delta E = |e|\, V_{bi}$$

with:

$$V_{bi} = \frac{k\,T}{|e|} ln\left(\frac{N_A\, N_D}{n_i^2}\right) \qquad [1.16]$$

where V_{bi} is the built-in potential of the junction. It represents the potential barrier that the electrons of the CB of the N-type SC (or the holes of the VB of the P-type SC) must surmount to reach the corresponding allowed band of the other SC.

When the junction is biased by a voltage V_0, the potential barrier is changed. Under forward bias, it decreases from V_0 and becomes $V_{bi} - V_0$. As a result, the width of the SCR is reduced, and the current of the diode increases. Under reverse bias, the barrier increases from V_0, becoming $V_{bi} + V_0$. The width of the SCR increases and the current in the diode is practically zero.

1.3.2. *Junction capacitance*

The SCR of the junction, of permittivity $\varepsilon = \varepsilon_r\, \varepsilon_0$, contains charges of opposite signs of value Q. These charges increase in reverse bias of the junction, and the increase dQ is a function of the increase dV_0 of the applied voltage. The junction capacitance per unit area is defined by:

$$C_j = \frac{dQ}{dV_0} \qquad [1.17]$$

For the P–N junction, this capacitance is a function of the concentrations of impurities, following the expression:

$$C_J = \left[\frac{|e|\, \varepsilon_r\, \varepsilon_0\, N_A N_D}{2\,(V_{bi} + V_0)(N_A + N_D)}\right]^{1/2} \qquad [1.18]$$

1.4. Impurities and defects

A *point defect* is defined as a localized break in the periodicity of the crystal lattice. In SCs, even pure ones, the concentration of point defects is high. The methods used to synthesize the materials allow the introduction of many impurities into the structure of the crystals. These are *extrinsic defects* that are impurity atoms in interstitial or substitutional forms. Other imperfections in the lattice can be found in vacancy or interstitial impurities, which are called *intrinsic defects*.

Point defects play an essential role in the physical properties, including the electronic properties, of SCs. In fact, the periodic potential of the crystal lattice is disrupted by local breakages as a result of the defects that create additional energy levels, discrete or distributed, known as *trap levels* or *localized levels* in the SC gap. These levels change the mechanisms of the charge transport, recombination and optical transitions, and can alter the initial or expected properties of an SC.

In addition to point defects, there are also extended defects of one, two or three dimensions, such as *dislocations* (1D), *grain boundaries* (2D) or *clusters* (3D) that form systematically or randomly in the material, depending on the conditions of the fabrication, treatment or shaping.

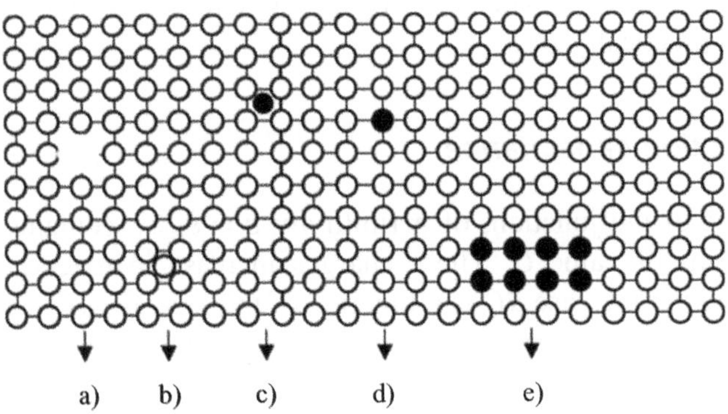

Figure 1.3. *Structural defects in semiconductors: a) vacancy; b) interstitial atom; c) interstitial impurity; d) substitutional impurity; e) substitutional impurity cluster*

1.4.1. *Traps and recombination centers*

The localized levels in SC gaps can interact with free carriers in the allowed bands via electrostatic interactions. When conditions are favorable, these defects capture the carriers and keep them trapped for a certain period of time. Then, the trapped carriers are released through an external energy supply (in the form of thermal, optical or electrical energy). Depending on the nature of the transitions between bands and localized levels, we can distinguish two types of centers: traps and recombination centers (see Figure 1.4).

1.4.1.1. *Traps*

A trap can capture a free carrier from an allowed band, hold it for a period of time and then release it. The trapped carrier returns to the allowed band where it was before. Traps can be electrons (from the CB) or holes (from the VB). Trapping involves a localized level and an allowed band (CB or VB).

1.4.1.2. *Recombination centers*

A recombination center can capture a free carrier of an allowed band, retain it for a period of time and then associate it with a charge carrier of the opposite charge. This recombination involves one localized level and two allowed bands (CB and VB).

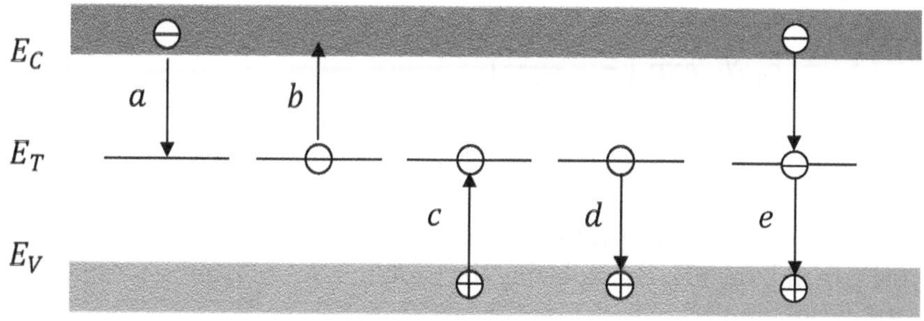

Figure 1.4. *Trapping and recombination mechanisms: a) capture of an electron; b) emission of an electron; c) capture of a hole; d) emission of a hole; e) recombination*

1.4.1.3. *Energy level of the trap center E_T*

The difference in energy $E_C - E_T$ or $E_T - E_V$ is the depth of the trap center or the activation energy of the trapped charge carriers.

1.4.1.4. *Capture rates c_n and c_p*

This is the number of electrons or holes captured per second (in s^{-1}). The capture rate of carriers is a function of the capture cross-section σ_n or σ_p of the defect. This parameter represents the surfaces surrounding the defect in which a moving charge carrier will be captured (in m^2). The expression of the carrier capture rate is given by:

$$c_n = \sigma_n \langle v_{thn} \rangle \qquad [1.19a]$$

$$c_p = \sigma_p \langle v_{thp} \rangle \qquad [1.19b]$$

where $\langle v_{thn} \rangle$ and $\langle v_{thp} \rangle$ are, respectively, the average thermal velocities of electrons and holes.

1.4.1.5. *Emission rates e_n and e_p*

This is the number of electrons or holes emitted per second (in s^{-1}).

At thermodynamic equilibrium, taking account of the processes of the emission and capture of carriers, we can obtain the expressions of the emission rates of electrons and holes:

$$e_n = c_n N_C \, exp\left(-\frac{E_C - E_T}{k\,T}\right) \quad [1.20a]$$

$$e_p = c_p N_V \, exp\left(-\frac{E_T - E_V}{k\,T}\right) \quad [1.20b]$$

where N_C and N_V are, respectively, the effective or equivalent concentrations of the CB and the VB. By replacing the capture rates of expressions [1.19] in [1.20], we obtain:

$$e_n = A\,T^2 \sigma_n \, exp\left(-\frac{E_C - E_T}{k\,T}\right) \quad [1.21a]$$

$$e_p = B\,T^2 \sigma_p \, exp\left(-\frac{E_T - E_V}{k\,T}\right) \quad [1.21b]$$

where A and B are constants independent of temperature.

Assuming that the capture cross-sections are temperature independent, plotting $ln\left(\frac{e_{n,p}}{T^2}\right)$ as a function of $\left(\frac{1}{T}\right)$ determines the defect level E_T. This method is commonly used in measuring trap parameters.

It should be noted that the energy level E_T rarely reflects a discrete level of defects, but more generally, the average level of a distribution of defects in the gap. Trap distributions can be exponential or Gaussian. The concentration of defects, N_T, which is the number of defects per unit volume of the material, can also be determined experimentally. The way in which it is determined depends on the technique used. Many methods can be used for measuring defect parameters. The most common and useful methods are electrical and optical measurements.

1.4.1.6. *Principle of electrical defect measurements*

For electrically measuring defects, all techniques have three steps:

1) the filling of localized states by charge carriers;

2) a change in the occupancy status of the defects;

3) a direct or indirect measurement of the occupancy modification of defects.

Several methods are used to study SCs, including thermally stimulated current (TSC), impedance spectroscopy (IS) and deep-level transient spectroscopy (DLTS). We will now give a description of the principle of the conventional DLTS method.

1.4.1.6.1. Deep-level transient spectroscopy method

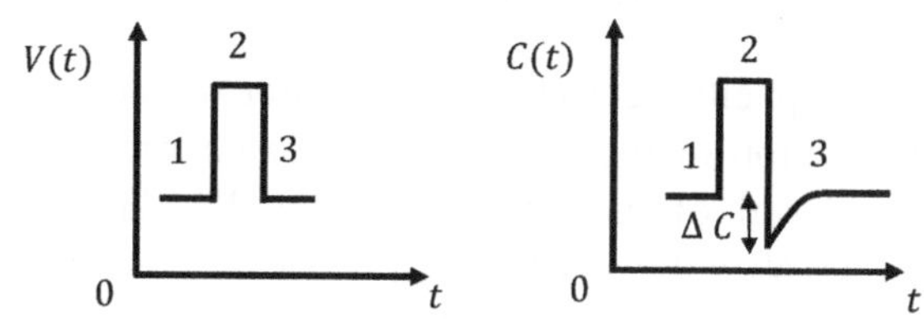

Figure 1.5. *The pulse applied to the junction and the variation of the junction's capacitance as a function of time*

A positive pulse is applied to the SC-based device in forward bias, in order to fill the defects. Then, the negative or zero pulse is applied to the device in reverse bias and the variations of the capacitance ΔC are measured as a function of time (see Figure 1.5).

Consider the case of an N-type SC containing electron traps of a concentration N_T and centered at the energy level E_T. This SC is in contact with another SC or a metal and an SCR of width λ is formed. Under a forward bias by a voltage pulse, electrons are injected into the SCR, whose width is reduced and a number of them are captured by traps with a rate equal to c_n. At the end of the pulse, the trapped electrons are released with an emission rate equal to e_n. The width $\lambda(t)$ of the SCR then increases with time and regains its initial value (see Figure 1.6).

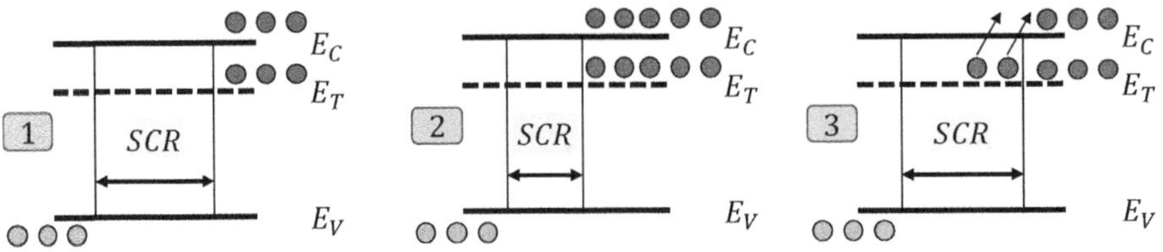

Figure 1.6. *Changes in the SCR of the junction during the trapping process: 1) before the application of the pulse; 2) trap filling; 3) trap release. For a color version of this figure, see www.iste.co.uk/nguyen/electronics1.zip*

Let S be the surface of the contact. The initial value of the junction capacitance is:

$$C_0 = \varepsilon_r \, \varepsilon_0 \frac{S}{\lambda} = S \left[\frac{|e| \, \varepsilon_r \, \varepsilon_0 \, N_D}{2 \, (V_{bi} + V_0 \,)}\right]^{1/2}$$

At time t, the capacitance $C(t)$ is a function of the width $\lambda(t)$ of the SCR, that is, a function of the concentration of charges released from the traps after the removal of the applied voltage V_0. The concentration of released charges is $N_T \, e^{-t/\tau_n}$, where $\tau_n = (e_n)^{-1}$ is the inverse of the emission rate of the trapped electrons. We obtain:

$$C(t) = S \left[\frac{|e| \, \varepsilon_r \, \varepsilon_0 \left(N_D - N_T \, e^{-t/\tau_n}\right)}{2 \, (V_{bi} + V_0 \,)}\right]^{1/2}$$

$$C(t) = S \left[\frac{|e| \, \varepsilon_r \, \varepsilon_0 \, N_D}{2 \, (V_{bi} \, + \, V_0 \,)}\right]^{\frac{1}{2}} \times \left[1 - \left(\frac{N_T}{N_D}\right) e^{-\frac{t}{\tau_n}}\right]^{\frac{1}{2}}$$

$$= C_0 \left[1 - \left(\frac{N_T}{N_D}\right) e^{-t/\tau_n}\right]^{1/2}$$

If $N_T \ll N_D$, we can write:

$$C(t) \sim C_0 \left[\left(1 - \left(\frac{N_T}{2N_D}\right) e^{-t/\tau_n}\right)\right] \qquad [1.22]$$

The measurement of changes in transient capacitance allows access to the trap concentration and the emission rate of trapped carriers.

We will use a variable called the *time window*, τ, defined by two times, t_1 and t_2, as:

$$\tau = \frac{t_2 - t_1}{ln(t_2/t_1)} \qquad [1.23]$$

The choice of the time window, and thus the times t_1 and t_2, is essential to obtain a usable spectrum of the curve $\Delta C(\tau)$. For a given temperature, the curve $\Delta C(\tau)$ shows a maximum for which the time window τ_M corresponds to the inverse of the emission rate of captured carriers. For electron traps, we obtain:

$$\tau_M = e_n^{-1} \qquad [1.24]$$

1.4.1.6.2. Determination of the defect parameters

DLTS measurements are performed by varying the temperature of the device containing traps, which are assumed to be electron traps. Using expression [1.24], the values of the emission rates e_n are determined. The plot of $ln\left(\frac{e_n}{T^2}\right)$ versus $\left(\frac{1}{T}\right)$ is normally a straight line according to the expression [1.21]. The slope of the line provides the activation energy E_T, and the capture cross-section can then be calculated by the intercept of the straight line with the Y-axis.

1.4.1.6.3. Variations of the deep-level transient spectroscopy method

The conventional DTLS method measures the capacitance variations of the device as a function of time. Depending on the structure of the sample and its dimensions, and also on the performance of measuring devices, the variations $\Delta C(\tau)$ can be small and impossible to perceive. The substitution methods measure changes in charge or current intensity (Q-DLTS or I-DLTS) and are effective for components such as organic SCs.

1.4.1.7. *Principle of optical defect measurements*

The defects contained in SCs can affect the luminescence of the material, varying to a greater or lesser extent depending on their energy position in the gap. To detect them, *photoluminescence spectroscopy* (PL) is frequently used. In this process, electron–hole pairs or excitons are created by optical excitation from a light source of an energy greater than that of the SC gap E_G. The exciton migrates into the crystal until it recombines or meets a defect site. If the energy of the incident photon is much greater than the SC gap, the electrons are deeply injected into the CB. They then relax by emitting phonons (through the thermalization process) to the E_C level and can perform recombinations from this level. The SC then relaxes towards thermodynamic equilibrium and the electron–hole pairs recombine through different processes involving photons (*radiative recombinations*) or phonons (*non-radiative recombinations*).

Each process involving radiation corresponds to a determined level of energy, and therefore to a precise wavelength, and the measurement of this wavelength may allow us to determine the energy levels of the electrons or holes that are at the origin of light emission (see Figure 1.7). The radiative recombination includes two types of PL emission: intrinsic emission and extrinsic emission.

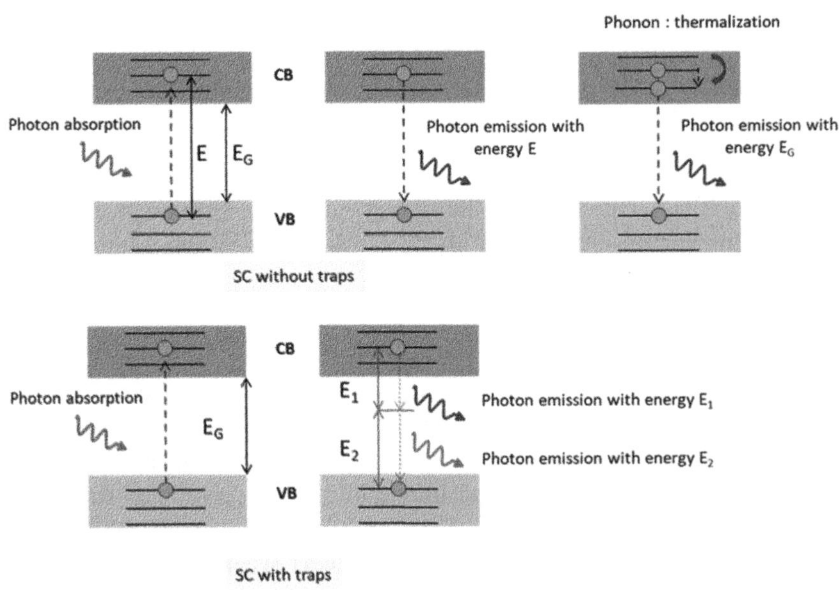

Figure 1.7. *The process of light emission in SCs, with and without defects. For a color version of this figure, see www.iste.co.uk/nguyen/electronics1.zip*

1.4.1.7.1. Intrinsic emission

This emission is produced by the recombination of electrons from the CB and holes from the VB. Since the binding energy of an exciton E_x is generally low (<20 meV), the energy released by an exciton during recombination is:

$$E_E = E_G - E_x \sim E_G = h\nu \qquad [1.25]$$

The energy of the emitted photon is close to that of the gap.

1.4.1.7.2. Extrinsic emission

Extrinsic emission is produced by localized charge carriers in the gap. These carriers are captured by the localized levels in the gap, created by the defects in the SC.

When in contact with defects, the binding energy of an exciton increases and it becomes bound with the defect. This is known as a bound exciton. The energy of the exciton includes the energy of the Coulombic interaction between the electron and the hole, and the potential energy created by the presence of the defect.

The recombination of the bound exciton occurs similar to that of a free exciton, but the energy of the photon emitted is different.

Let D be a donor atom and A an acceptor atom in the SC. To ionize these atoms, an energy of E_D must be supplied to atom D and an energy of E_A to atom A. These energies are called *ionization energies*.

The ionized atom D can trap an electron and the ionized atom A can trap a hole. The exciton formed by this electron and this hole can recombine, and the process is called *donor–acceptor recombination*.

The energy of the corresponding photon is:

$$E_E = E_G - (E_D - E_A) + \frac{e^2}{4\,\pi\,\varepsilon\,R} = h\nu_d \qquad [1.26]$$

where R is the distance between the two ionized atoms, $\varepsilon = \varepsilon_r \times \varepsilon_0$ is the permittivity of the SC and ν_d is the recombination frequency.

If R is low (high density of defects), a series of discrete lines may be observed in the PL spectrum. If R is large (low density of defects), there will be a continuum of lines whose lower limit is:

$$E_{Em} = E_G - (E_D - E_A) = h\nu_{dm} \qquad [1.27]$$

1.5. Metal/semiconductor contact

For electrical connections to electronic devices, it is necessary to make metal/SC junctions or contacts by metallizing the surface of the SC. As with P–N junctions, metal/SC contacts are the location of special mechanisms that have a significant influence on the charge transport in the active SC.

1.5.1. *Parameters of metal/semiconductor contacts*

For studying metal/SC contacts, we will define physical parameters relating to the energy of electrons in materials.

The potential energy of an electron located in vacuum in the immediate vicinity of the solid is referred to as the vacuum level. To extract an electron from a metal, it must be provided with an amount of energy equal to the difference in energy between the vacuum level and the Fermi level E_{Fm} where the electron is located. This amount of energy is called the *work function* of the metal, which is denoted as $|e|\,\Psi_m$.

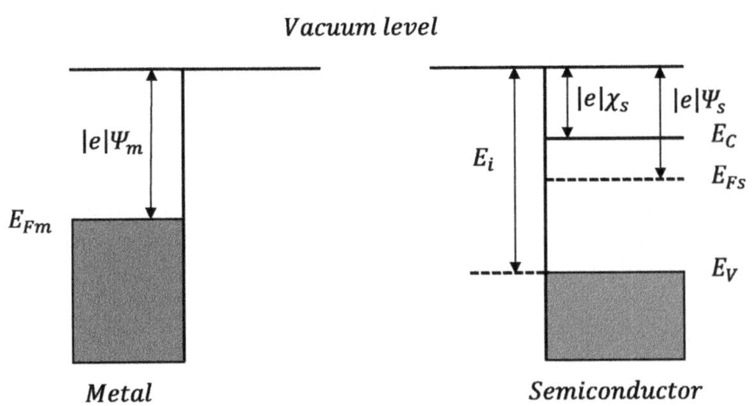

Figure 1.8. *The parameters in the energy band diagram for metals and SCs. For a color version of this figure, see www.iste.co.uk/nguyen/electronics1.zip*

For SCs, the Fermi level E_{Fs} is not set in the gap, but instead depends on the doping, and the work function $|e|\Psi_s$ is thus not constant. Therefore, this parameter is not often used. We can define the two energy parameters relating to the allowed bands for SCs as follows:

– the ionization energy is the minimum energy necessary to extract an electron from the top of the VB and bring it to the vacuum level. It is denoted as E_i (or *IE*);

– the electron affinity is the maximum energy necessary to extract an electron from the bottom of the CB and bring it to the vacuum level. It is denoted as $|e|\chi_S$.

The different parameters of the metal/SC contact are shown in Figure 1.8.

1.5.2. *Formation of metal/semiconductor contacts*

When a metal and an SC are brought into contact, free carriers flow from one material to another, until equilibrium is reached. As in the case of the P–N junction, the alignment of the Fermi levels of materials is assumed and, consequently, a transfer of charges between two materials takes place, resulting in the formation of an SCR at the metal/SC interface. This region forms the contact potential barrier and prevents further transfer of charges in both directions if no electric field is applied.

Depending on the energy characteristics of the energy bands of materials, there may be several possible types of electrical contacts. Since the position of the Fermi level in the gap of a doped SC differs depending on the type of doping, we must distinguish N-type and P-type SCs. Let us consider the case of an N-type SC.

1.5.2.1. *Neutral contacts*

These contacts fulfill the condition that the regions adjacent to the physical contact should be electrically neutral. For this contact, the work function of the metal and the SC is equal and no charge is transferred between the materials, nor are SCRs formed at the interface.

1.5.2.2. *Blocking contacts*

These contacts respond to the condition that $|e|\Psi_m > |e|\Psi_s$. In this case, the electrons flow from the SC to the metal, leading to the formation of a positive SCR due to the charge of the donor centers that are no longer compensated by the electronic charge. The alignment of the Fermi levels leads to an upward band-bending. When an equilibrium is reached, a potential barrier is formed at the interface and the electrons from the metal must cross this barrier to enter the SC.

1.5.2.3. *Ohmic contacts*

These contacts respond to the condition that $|e|\Psi_m < |e|\Psi_s$. In this case, the electrons of the metal are injected into the SC, forming a negative SCR in the vicinity of the contact. The alignment of the Fermi levels leads to a downward band bending.

The contact types mentioned previously are also found in P-type SCs but the effects are symmetrical. Figure 1.9 summarizes the different types of metal/SC contacts.

For the blocking contacts between a metal and an N-type SC, the height of the potential barrier is:

$$|e|\,\Phi_{bn} = |e|\,(\Psi_m - \chi) \tag{1.28}$$

For the blocking contacts between a metal and a P-type SC, the height of the potential barrier becomes:

$$|e|\,\Phi_{bp} = E_G - |e|\,(\Psi_m - \chi) \tag{1.29}$$

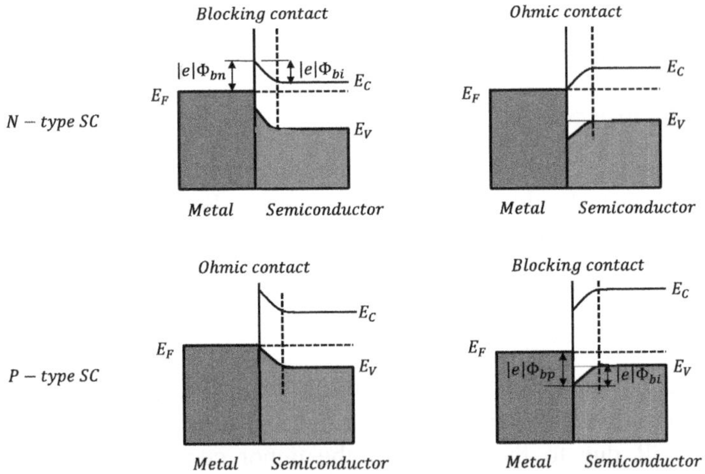

Figure 1.9. *The parameters in the energy band diagram for metals and SCs. For a color version of this figure, see www.iste.co.uk/nguyen/electronics1.zip*

Aligning the Fermi levels also allows the creation of a diffusion potential or a built-in potential V_{bi}, such that $|e|V_{bi}$ is equal to the difference between the work function:

$$|e|V_{bi} = \big||e|\,(\Psi_m - \Psi_s)\big| = |e| \times |(\Psi_m - \Psi_s)| \qquad [1.30]$$

1.5.3. *Width λ of the space charge region*

The width of the SCR formed at the metal/SC contact will depend on the potential V_{bi}. Using Poisson's equation and expression $\vec{E} = -\overrightarrow{grad}\,V = -\vec{\nabla}\,V$, the following relations are established between the diffusion potential and the width of the SCR for an N-type SC:

$$V_{bin} = \frac{|e|\,N_D\,\lambda^2}{2\,\varepsilon_r\,\varepsilon_0} \qquad [1.31]$$

$$\lambda = \left(\frac{2\,\varepsilon_r\,\varepsilon_0\,V_{bin}}{|e|\,N_D}\right)^{1/2} \qquad [1.32]$$

When a voltage V_0 is applied to the junction, the potential V_{bin} is changed. In forward bias ($V_m > V_S$), the applied voltage decreases the total potential and causes a narrowing of the SCR, which becomes:

$$\lambda = \left[\frac{2\,\varepsilon_r\,\varepsilon_0(\,V_{bin} - V_0)}{|e|\,N_D}\right]^{1/2} \qquad [1.33]$$

Similar to a P–N junction, reverse bias increases the built-in potential and width of the SCR.

1.5.4. *Junction capacitance*

The SCR of the metal/SC contact is a capacitor with a capacitance of C_{SC}. The charge of the capacitor is due to the ionized impurities of the SC. For a forward-biased junction through V_0, the density of the charge per unit area is:

$$Q_{SC} = |e|\,N_D\lambda = [2\,|e|\varepsilon_r\,\varepsilon_0\,N_D(\,V_{bin} - V_0)]^{1/2} \qquad [1.34]$$

The junction capacitance per unit area of the corresponding SC under forward bias is:

$$c_{SC} = \frac{dQ_{SC}}{dV_0} = \left(\frac{|e|\varepsilon_r\,\varepsilon_0\,N_D}{2\,(V_{bin} - V_0)}\right)^{1/2} \quad [1.35]$$

1.5.5. *Schottky effect*

The Schottky effect refers to the lowering of the potential barrier at the metal/SC contact due to the image force when applying an electric field $\vec{E}$ to the interface.

This force originates from the charge induction of the surface of the metal that is in contact with the SC and exerts a force equal to the force of a symmetrical charge of the electron (known as the *image charge*; see Figure 1.10).

Let x be the distance of an electron from the interface. The image force exerted on this electron is:

$$F(x) = -\frac{1}{4\,\pi\,\varepsilon_r\varepsilon_0} \times \frac{|e|^2}{(2\,x)^2} \quad [1.36]$$

The potential energy associated with the image force is:

$$E(x) = \int_{-\infty}^{x} F(x) = -\frac{|e|^2}{16\,\pi\,\varepsilon_r\varepsilon_0 x} \quad [1.37]$$

The total potential energy of the electron when an electric field is applied to the junction is:

$$E_P(x) = -\frac{|e|^2}{16\,\pi\,\varepsilon_r\varepsilon_0 x} - |e|Ex \quad [1.38]$$

The maximum value of the function $E_P(x)$ is obtained for:

$$x_m = \left(\frac{|e|}{16\,\pi\,\varepsilon_r\varepsilon_0\,E}\right)^{1/2} \quad [1.39]$$

We can find the barrier lowering as a result of the image force and in the presence of an electric field $\vec{E}$:

$$\Delta\Phi = \left[\frac{(|e|)^3\, E}{4\,\pi\,\varepsilon_r\varepsilon_0}\right]^{1/2} \quad [1.40]$$

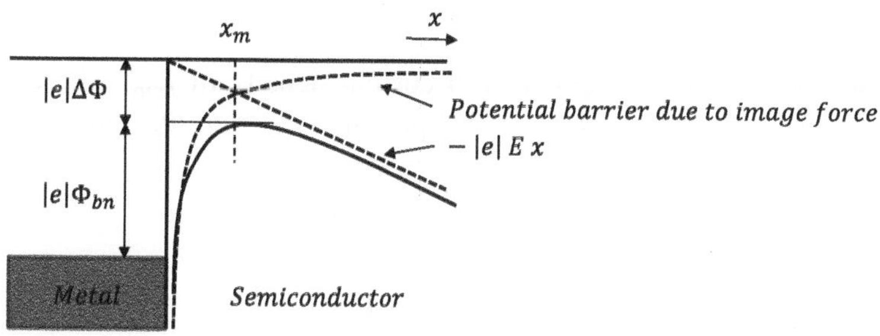

Figure 1.10. *Schottky barrier at a metal/SC interface with the Schottky effect*

1.5.6. *Schottky diode*

When a voltage is applied to the metal/SC contact, a current flows between the metal and the SC. Consider the case of an N-type SC that forms a blocking contact with the metal (see Figure 1.8). As with the case of the P–N junction, the forward bias (when the potential of the metal is greater than the potential of the SC, $V_m > V_S$) decreases the built-in V_{bi} and causes the potential barrier to be lowered by $\Delta\Phi$. The current flow from the SC to the metal is made easier and the density of the current is high.

On the other hand, the reverse bias ($V_m < V_S$) increases V_{bi} and the potential barrier $\Delta\Phi_{bn}$ is greater. As a result, the current density decreases. The metal/SC junction therefore behaves like a junction diode, which is called a *Schottky diode*.

For ohmic contacts, charge carriers do not cross a potential barrier, and theoretically, the current density increases linearly with the applied voltage. However, the presence of traps will change the charge transport in the devices.

Unlike the P–N junction, the current flowing through the metal/SC junction is provided by the majority carriers. In the case of an N-type doped SC, the electrons of the CB cross the potential barrier by various processes, depending on the characteristics and conditions of the junction. The mechanism that is often used in describing the transport of electrons is thermionic emission, while other possible mechanisms are the tunnel effect and the thermally assisted tunnel effect.

In thermionic emissions, the electrons move under forward bias V_0 in two directions, as shown in Figure 1.11:

– from the SC to the metal with a current density of J_{sm}. The electrons with sufficient energy pass over the potential barrier Φ_{bn} and enter the metal;

– from the metal to the SC with a density of J_{ms}.

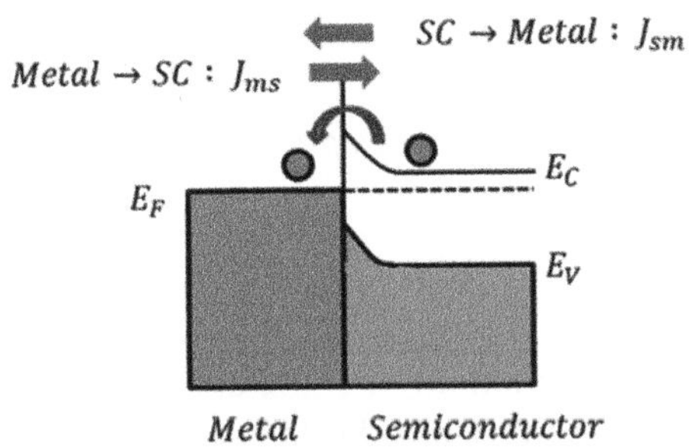

Figure 1.11. *Electron transport in a contact between a metal and an N-type SC. For a color version of this figure, see www.iste.co.uk/nguyen/electronics1.zip*

The resulting current density is $J_{SC} = J_{sm} - J_{ms}$. It will depend on the potential barrier and the applied voltage, following the expression:

$$J_{SC} = \left[A^* T^2 \, exp\left(-\frac{|e|\Phi_{bn}}{k\,T}\right)\right]\left[exp\left(\frac{|e|\,V_0}{k\,T} - 1\right)\right] \qquad [1.41]$$

where:

$$A^* = \frac{4\,\pi\,|e|\,m_n^*\,k^2}{h^3} \qquad [1.42]$$

The parameter A^* is the effective *Richardson constant*. The saturation current density is defined by:

$$J_s = A^* T^2 \, exp\left(-\frac{|e|\Phi_{bn}}{k\,T}\right) \qquad [1.43]$$

The expression for the total current density thus becomes:

$$J_{sc} = J_s \left[exp\left(\frac{|e|\, V_0}{k\, T}\right) - 1 \right] \qquad [1.44]$$

This expression is similar to the expression for the current in a P–N junction under forward bias. The characteristic current–voltage density shows exponential growth under forward bias, and little to no change in reverse bias. However, the transport mechanisms in the junctions are different. In the case of the Schottky junction, charge carriers, which are majority carriers, cross a potential barrier, while in the case of the P–N junction, the current occurs due to the diffusion of the minority charge carriers of the two SCs.

For an accurate calculation, the reduction $|e|\Delta\Phi$ of the barrier by the Schottky effect must be considered, and the expression of the potential barrier must be corrected by replacing $|e|\Phi_{bn}$ with $|e|\Phi_{bn} - |e|\,\Delta\Phi$. Under these conditions, the expression [1.43] becomes:

$$J_s = A^* T^2\, exp\left(-\frac{|e|\Phi_{bn}}{k\, T}\right) exp\left(\frac{|e|\Delta\Phi}{k\, T}\right) \qquad [1.45]$$

The different conduction mechanisms that may occur during contact between metals and organic/inorganic SCs are discussed in Chapters 4 and 5.

1.6. Semiconductors under non-equilibrium conditions

For an SC in thermal equilibrium, the mass action law applies, and we obtain relationship [1.4] between the concentrations of electrons n_0 and the holes p_0: $n_0 \times p_0 = n_i^2$. If this equilibrium is disturbed by an energy supply from an external source, new charge carriers are created (known as *excess carriers*) and are added to existing charges. The mass action law no longer holds true until the SC returns to equilibrium.

1.6.1. *Parameters of a semiconductor under non-equilibrium conditions*

The *rate of thermal generation* $G_{n,p}$ is defined as the number of carriers created per unit of volume and per unit of time (in $cm^{-3}s^{-1}$). Similarly, the *recombination rate* $R_{n,p}$ is the number of carriers that disappear by recombination per unit of volume and per unit of time (in $cm^{-3}s^{-1}$). Since the processes of generation and recombination involve electron–hole pairs, we can make considerations about only one type of carrier, such as electrons.

The concentrations of electrons and holes are defined by the following parameters:

– n_0 p_0: concentrations of electrons and holes at thermal equilibrium (at a constant temperature);

– n, p: total concentrations of electrons and holes.

Let dn and dp be the excess concentrations of electrons and holes. We obtain:

$$dn = n - n_0 \quad [1.46a]$$

$$dp = p - p_0 \quad [1.46b]$$

In the case of a direct electron–hole recombination and low injection conditions ($dn = dp \ll n_0$ or p_0), the recombination rate is proportional to the concentrations of electrons and holes, that is:

$$R_n = R_p = B\, n\, p \quad [1.47]$$

where B is called the bimolecular recombination rate.

For an SC at thermal equilibrium, equation [1.47] can be written as:

$$R_{n0} = G_{n0} = B\, n_0\, p_0$$

The recombination rate of excess electrons is therefore:

$$r_n = -\frac{d[dn(t)]}{dt} = (R_n - R_{n0}) = B\,(n\,p - n_0\,p_0)$$

$$r_n = -\frac{d[dn(t)]}{dt} = B\left((n_0 + dn)\,(p_0 + dp) - n_0\,p_0\right)$$

$$r_n = B(n_0\,dp + p_0\,dn + dn\,dp)$$

For a P-type SC, we obtain $p_0 \gg n_0$ and the previous equation becomes:

$$\frac{d[dn(t)]}{dt} = -B\,p_0\,dn$$

That is:

$$dn(t) = dn(0)e^{-1/\tau_n}$$

with:

$$\tau_n = \frac{1}{B\,p_0} \qquad [1.48]$$

where τ_n is called the *lifetime of excess minority electrons*. Since the majority holes recombine with the minority electrons at the same rate, we obtain:

$$r_p = r_n = -\frac{d[dn(t)]}{dt} = \frac{dn(t)}{\tau_n} \qquad [1.49]$$

For an N-type SC, the lifetime of the excess minority holes is:

$$\tau_p = \frac{1}{B\,n_0} \qquad [1.50]$$

The recombination rate of minority holes is equal to the recombination rate of majority electrons:

$$r_n = r_p = -\frac{d[dp(t)]}{dt} = \frac{dp(t)}{\tau_p} \qquad [1.51]$$

1.6.2. *Recombination of carriers via recombination centers (Shockley–Read–Hall theory)*

Consider an SC with recombination centers located at the energy level E_T with a concentration N_T. Taking into account the processes of capturing (of coefficients c_n and c_p) and of emission (of coefficients $e_n$and e_p), we obtain the following expression for the recombination rates of electrons and holes:

$$r = r_n = r_p = \frac{c_n\,c_p\,N_T(n\,p - n_i^2)}{c_n(n + n\prime)+c_p(p + p\prime)} \qquad [1.52]$$

where:

$$n' = N_C \, exp\left[-\frac{(E_C - E_T)}{k\,T}\right] \qquad [1.53a]$$

$$p' = N_V \, exp\left[-\frac{(E_T - E_V)}{k\,T}\right] \qquad [1.53b]$$

If the recombination center is close to the middle of the gap and the capture coefficients are such that $c_n \sim c_p = c$, expression [1.52] can be simplified to:

$$r = r_n = r_p = \frac{1}{\tau_R}\frac{n\,p - n_i^2}{2\,n_i + p + n} \qquad [1.54]$$

where:

$$\tau_R = \frac{1}{c\,N_T} \qquad [1.55]$$

For an N-type SC, we obtain:

$n = n_0 + dn \sim n_0$ and $p = p_0 + dp \sim p_0$

Expression [1.54] can be simplified to:

$$r = r_R = \frac{1}{\tau_R}\frac{n\,p - n_i^2}{2\,n_i + p + n} = \frac{1}{\tau_R}\frac{n\,p - n_i^2}{n} = \frac{p - p_0}{\tau_R} = \frac{dp}{\tau_R}$$

The recombination process via recombination centers is added to the direct recombination process and the total recombination rate in an N-type SC becomes:

$$r_R = \frac{dp}{\tau_p} + \frac{dp}{\tau_R} = \frac{dp}{\tau_T} \qquad [1.56]$$

with:

$$\frac{1}{\tau_T} = \frac{1}{\tau_p} + \frac{1}{\tau_R} = Bn_0 + cN_T \qquad [1.57]$$

1.6.3. *Transient relaxation current*

An SC out of equilibrium returns to its original state once the disturbance created by the external source has been removed. This return to equilibrium is not instantaneous, and the changes in the concentrations of carriers create a transient relaxation current reflecting the decay over time of the excess concentration of carriers.

Consider an N-type SC uniformly illuminated by a light source that creates excess carriers with a generation rate of $g_L = g_p = g_n$.

In a low excitation regime, the recombination rate of minority carriers (holes) is given by expression [1.51]:

$$r_p = \frac{dp(t)}{\tau_p} = \frac{p - p_0}{\tau_p} \quad [1.58]$$

The change in the concentration of holes is equal to the difference between the generation and recombination rates. There are several such cases that can be examined.

In the steady state, that is, when the SC is in equilibrium while illuminated, we obtain:

$$\frac{dp(t)}{dt} = 0 = g_p - \frac{p - p_0}{\tau_p} \quad [1.59]$$

From which:

$$g_p = \frac{p - p_0}{\tau_p}$$

We then deduce:

$$p - p_0 = g_p \tau_p \quad [1.60]$$

The concentration of excess minority carriers is equal to the product of the generation rate and the lifetime of these carriers.

In the transient regime, that is, when the SC has reached the equilibrium regime under illumination and the light source is removed, the generation rate is zero and equation [1.59] becomes:

$$\frac{dp(t)}{dt} = \frac{p - p_0}{\tau_p} \qquad [1.61]$$

The boundary conditions of the process of decay of excess concentration are:

– $t = 0: p(0) = p_0 + g_p \tau_p$;

– $t \to \infty: p(\infty) = p_0$.

The change in the concentration of excess carriers as a function of time is given by:

$$p(t) = p_0 + g_p \tau_p \, exp\left(-\frac{t}{\tau_p}\right) \qquad [1.62]$$

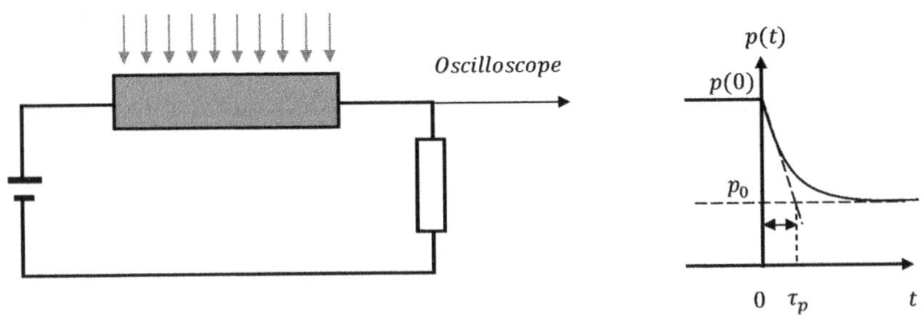

Figure 1.12. *The change in the concentration of excess carriers as a function of time*

This equation is used to determine the lifetime of excess carriers using the set-up shown in Figure 1.12. The lifetime of the holes is equal to the time constant of the curve $p(t)$.

In the case of non-uniform *illumination*, that is, when only one part of the SC is illuminated, the excess concentration of carriers is localized within this part and the diffusion of carriers will occur to balance the carrier distribution (see Figure 1.13). We will assume that the illumination is permanent on one side of the SC. By writing the continuity equation, which expresses that the change in the concentration of carriers is equal to the difference between the generation rate and the recombination rate, to which we add the difference between the output current and the input current, the following expression is obtained for the diffusion in the direction of the output current Ox:

$$\frac{\delta p_n}{\delta t} = 0 = -\frac{p_n - p_{no}}{\tau_p} + D_p \frac{\delta^2 p_n}{\delta x^2} \quad [1.63]$$

The boundary conditions of the process of decay of excess concentration are:

$$x = 0 : p(0) \; x \to \infty : p(\infty) = p_0$$

The change in the concentration of excess carriers as a function of time is given by:

$$p(x) = p_0 + (p(0) - p_0) \; exp\left[-\frac{x}{(D_p \times \tau_p)^{1/2}}\right]$$

$$p(x) = p_0 + (p(0) - p_0) \; exp\left(-\frac{x}{L_p}\right) \quad [1.64]$$

The diffusion parameter L_p for holes (and L_n for electrons) is called the *diffusion length,* which is given by:

$$L_p = \sqrt{D_p \tau_p} \quad [1.65a]$$

$$L_n = \sqrt{D_n \tau_n} \quad [1.65b]$$

The diffusion length of the carriers represents the diffusion distance of the carriers before they disappear through recombination.

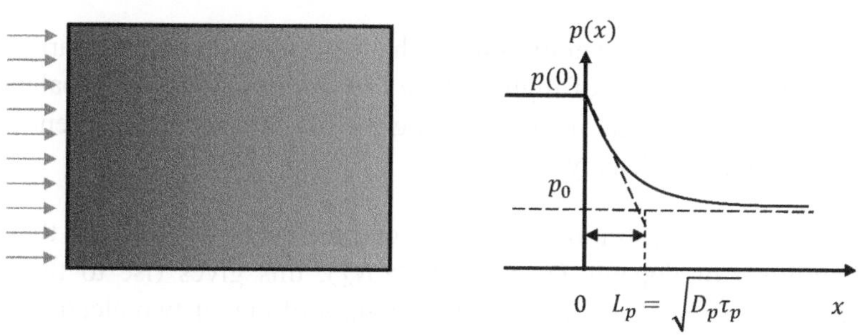

Figure 1.13. *Diffusion of excess carriers within the thickness of an SC lit on one side*

1.7. Space charge current

In an N-type SC in thermal equilibrium, the charge density ρ_{CE} is a function of the concentrations of dopants and the concentrations of free charges:

$$\rho_{CE} = |e|\,[n - p + N_A - N_D] \qquad [1.66]$$

Once the SC is in equilibrium, the concentrations of free carriers n and p are equal to the concentrations of dopants N_D and N_A. Therefore, ρ_{CE} is equal to zero.

For low applied voltages, the number of injected carriers is less than the number of thermally generated free carriers, and the current density is proportional to the applied voltage generated following Ohm's law. An average field is defined by the expression:

$$E = \frac{V}{d}$$

where d is the thickness of the sample.

The expression for the total current density for an N-type SC is thus written as:

$$J = |e|\,n\,v = |e|\,\mu_n n\,E$$

$$J = |e|\,\mu_n n\,E = |e|\,\mu_n n\,\frac{V}{d} \qquad [1.67]$$

The density of the current varies linearly with the applied voltage and the conduction regime is ohmic.

If the SC has a low level of doping, the concentrations of free carriers are low, and the application of high voltage can easily lead to an imbalance of charges. The free carriers injected during the bias change the concentrations n and p and thus the density ρ_{CE}.

When the concentrations of free carriers become greater than the concentrations of dopants ($n > N_D, p > N_A$), this gives rise to the space charge effect: an SCR is formed in the vicinity of one or two electrodes due to the fact that free charges can no longer be completely discharged from the volume of the SC.

As a result, the applied voltage drops, mainly in the SCR, and the electric field will no longer be uniform throughout the width of the SC.

The current that flows through the SC, in this case, is called space-charge limited current (SCLC).

1.7.1. *The case of an ideal semiconductor*

1.7.1.1. *A semiconductor without traps*

If the SCR does not contain traps, and if we do not consider the diffusion current along with the drift current, for an N-type SC subject to an electric field, Ohm's law is written as:

$$J = -|e|\, n\, v = -\,|e|\, \mu_n n\, E \qquad [1.68]$$

By applying Poisson's equation, we obtain:

$$\frac{d^2V}{dx^2} = -\frac{dE}{dx} = \frac{|e|\, n}{\varepsilon} \qquad [1.69]$$

From expressions [1.68] and [1.69], we can deduce:

$$\frac{J}{|e|\, \mu_n E} = n = \frac{\varepsilon\, dE}{dx}$$

As $E\, dE = \frac{J}{\varepsilon\, \mu_n}\, dx$:

$$\frac{1}{2} E(x)^2 = \int \frac{J}{\varepsilon\, \mu_n}\, dx = \frac{J\, x}{\varepsilon\, \mu_n} + A \qquad [1.70]$$

where A is a constant that cancels with the boundary conditions $E\,(x = 0) = 0$.

From which:

$$E(x) = \left(\frac{2\, Jx}{\varepsilon\, \mu_n}\right)^{1/2}$$

The potential related to the electric field is:

$$dV(x) = -E(x)\,dx$$

With d being the width of the SC, the voltage applied to the terminals of the sample will be:

$$V = \int_0^d \left(\frac{2\,J}{\varepsilon\,\mu_n}\right)^{1/2} x^{1/2}\,dx = \left(\frac{2\,J}{\varepsilon\,\mu_n}\right)^{1/2} \times \frac{2}{3}\,d^{3/2}$$

We then deduce:

$$J = \frac{9}{8}\,\varepsilon\,\mu_n \frac{V^2}{d^3} \qquad [1.71]$$

The current density limited by the space charge follows the expression [1.71], known as the Mott–Gurney law. This conduction regime is known as a trap-free regime.

The transition from the ohmic conduction regime to the trap-free conduction regime takes place for a voltage V_Ω such as:

$$J = |e|\,\mu_n n\,\frac{V}{d} = \frac{9}{8}\,\varepsilon\,\mu_n \frac{V^2}{d^3}$$

We then deduce:

$$V = V_\Omega = \frac{8}{9}\,\frac{|e|\,n\,d^2}{\varepsilon} \qquad [1.72]$$

1.7.1.2. *A semiconductor containing traps*

The presence of traps has a reduction effect on the SCLC current because most of the free carriers will be captured by the traps. In the case of electrons, the Mott–Gurney law becomes:

$$J = \frac{9}{8}\,\varepsilon\,\theta\,\mu_n \frac{V^2}{d^3} \qquad [1.73]$$

The parameter θ represents the ratio of the number of free charge carriers to the total number of charge carriers, that is, the sum of free charges and trapped charges. Thus, we obtain:

$$\theta = \frac{n_{free}}{n_{total}} = \frac{n_{free}}{n_{free}+n_{trapped}} \qquad [1.74]$$

In other words, the mobility of the carriers is reduced by the parameter θ (< 1) because the traps are present. The effective mobility of the carriers will be $\mu_{eff} = \theta\, \mu_n$. This can be determined experimentally from the current–voltage density curve using expression [1.72].

This conduction regime is known as a trap-filled regime.

1.7.2. *Trap-filled limit voltage*

In an SC that contains traps, a change in the electronic conductivity regime can be observed when all traps are filled.

Effectively, in this case, all the carriers that are injected are free to move, and the current increases with the voltage. In other words, when $\theta = 1$, the current density once again obeys the Mott–Gurney law.

The transition voltage between the two regimes is known as the *trap-filled limit voltage,* which is denoted as V_{TFL} (where TFL refers to trap-filled limit).

For shallow traps, the voltage V_{TFL} is:

$$V_{TFL} = \frac{|e|\, N_T\, d^2}{2\, \varepsilon} \qquad [1.75]$$

1.7.3. *Discrete traps and trap distribution*

Trap centers are defined as discrete energy levels in the SC gap, with an activation energy E_T and a concentration N_T. In reality, traps are distributed in terms of energy and concentration in most cases. The possible distribution types are exponential distributions and Gaussian distributions.

For an exponential distribution, the concentration of traps varies as a function of their energy level, which is given by:

$$n_T = \frac{N_T}{k\, T_C} \times exp\left(-\frac{E_C - E_T}{k\, T_C}\right) \qquad [1.76]$$

where T_C is the characteristic temperature of the distribution of the traps. We define the parameter m associated with this temperature as $m = \frac{T_C}{T}$. The relationship between the current density and voltage in this case is written as:

$$J = N_C\, \mu_n\, |e|^{1-m} \left[\frac{m\,\varepsilon}{N_T(1+m)}\right]^m \left(\frac{1+2m}{1+m}\right)^{1+m} \frac{V^{1+m}}{d^{1+2m}} \qquad [1.77]$$

The current density varies as a function of the voltage as determined by the power law $J \propto V^n$ $(n \neq 2)$.

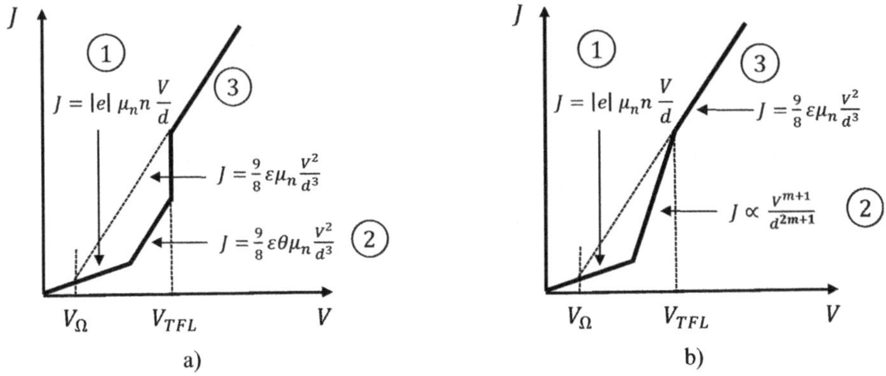

Figure 1.14. *Electronic conduction regime of the space-charge limited current. a) SC with a discrete trap level: (1) ohmic regime, (2) trap-filled regime, (3) trap-free regime. b) SC with an exponential distribution of traps: (1) ohmic regime, (2) trap-filled regime, (3) trap-free regime*

The different conductivity regimes of the current limited by the space charge are shown in Figure 1.14.

1.8. Hopping conduction

An SC can contain numerous defects that are active traps in the gap of an SC. Each individual defect corresponds to an energy level referred to as a localized level. The distribution of localized levels in the gap of an SC can be uniform, exponential, Gaussian or random. When the temperature of the material decreases, the number and energy of the phonons decreases, and the carrier emission to the allowed bands also decreases.

For localized levels close to the allowed bands, the charge carriers of localized levels can be emitted into these bands through a sufficient supply of energy. For deeper levels $E < E_F$, the carriers can move through jumps from one energy level to another energy level that is close by and located above the Fermi level with a low thermal energy supply of heat. This process is called hopping conduction.

It has been demonstrated that, at low temperatures, the process of hopping between localized states does not occur between the nearest localized states but between localized states with the smallest energy differences. This process is called variable-range hopping (VRH). The conductivity of these SCs (heavily doped or amorphous) depends on the temperature and follows the Mott–Davis law:

$$\sigma = A\,exp\left(-\frac{B}{T^{1/4}}\right) \qquad [1.78]$$

where A and B are characteristic constants of the material.

2

Materials

2.1. Introduction

There are many materials used in organic electronics and these materials differ from those of inorganic electronics in their physical characteristics and the role they play in components. Consider, for example, the basic structure of a light-emitting diode, which is a component used in technology for displays and lighting, and which can be manufactured with inorganic or organic semiconductors (SC) (see Figure 2.1).

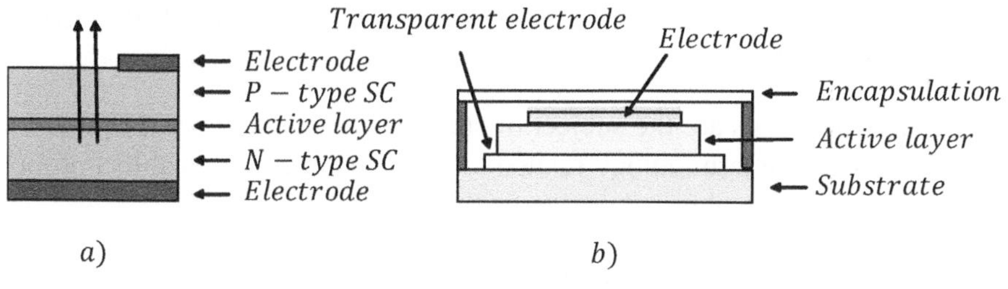

Figure 2.1. *Basic structure of an SC-based light-emitting diode: (a) inorganic SC; (b) organic SC*

The inorganic diode is formed by a P–N junction provided with two metal electrodes for electrical connections. The light emitted by the diode emerges into the external medium through the surface of an SC, which can be protected by an encapsulation (epoxy dome). For an organic diode, the emitting layer is an organic SC inserted between two electrodes, including one transparent electrode, which enables photons to be emitted into the external medium.

The diode is protected by an encapsulation layer that protects the active layer from contact with the external medium. This layer is attached to the substrate using an adhesive. To take advantage of organic components, which offer the benefits of lightness, flexibility and a large active surface, the layers forming the diode must be thin and flexible. On the other hand, it is necessary to use transparent and conductive materials for optoelectronic components (light-emitting diodes or solar cells). In this chapter, we will look at the different materials used in organic electronic components, along with their physical characteristics and how they are implanted to create the different parts of an organic electronic device.

2.2. Organic materials

As in traditional electronics, materials play a decisive role in the performance of electronic devices made with active organic materials. Although the basic materials are organic, component manufacturing often uses diverse and varied materials to reinforce, improve or modify the original characteristics and properties of both the active material and the structure of the device in order to achieve the best possible performance. This chapter first examines the organic materials that are the essential elements of the components and then the organic and/or inorganic materials that enter the composition of the materials and the structure of the organic electronic devices.

Organic materials are the subject of study in this work and it will thus be useful to recall the terminology related to these materials, to enable a proper understanding of the physical processes that we will describe.

Overall, the term *organic* refers to materials containing carbon (C). They are often found in the form of compounds, which contain carbon atoms and other atoms such as hydrogen (H), oxygen (O), nitrogen (N), sulfur (S) or phosphorus (P). Atoms are bound together by covalent bonds, forming *molecules* or *repeating patterns*. Such patterns are called *constituent units* or groups of atoms (and are sometimes incorrectly referred to as *monomers*). The shape of the molecules can be linear or may consist of a complex structure.

Based on the size of the molecules of the material, we can distinguish two types of organic materials: small molecules and polymers. Small molecules are materials made up of groups of atoms, and their molar mass or molecular weight is usually low ($< 2{,}000\,\text{g.mol}^{-1}$). Polymers consist of chains of polymers with high molar masses ($> 20{,}000\,\text{g.mol}^{-1}$). The number of chains they contain is generally high (several thousands or more). Polymers with short chains or a small number of monomers (less than a few dozen) are called *oligomers*. The structure of the organic material influences its physical properties and must be taken into account in how it is prepared and shaped. In general, both small molecules and oligomers have low solubility in conventional solvents, and therefore vacuum evaporation is required to deposit films from these materials. Conversely, polymers can be dissolved in appropriate solvents and then deposited on substrates using various different techniques to create thin films.

Homopolymers differ from *copolymers* in the nature of the atoms that form their chains. In the former, the atoms are identical and the polymer is represented by $(A)_n$, where n is the number of molecules in the chain. In the latter, the molecules are of different types, and the polymer is represented by $-A-B-C-\ldots$. Depending on the polymerization conditions, the molecules of the copolymers can be randomly arranged, e.g., $-A-A-B-B-A-B-\ldots$, or in identical blocks, e.g., $-[A]_m-[B]_n-[A]_p-\ldots$. The parameter n is generally variable for polymers, which means that the lengths of the chains are not the same in a sample. They are statistically distributed and vary from one sample to another. To estimate the value of n, we determine the *number-average molar mass* (or molecular weight) $\langle M_n \rangle$ defined as the average of the masses of the species constituting the polymer per mole of material. If M_A is the molar mass of $-A-$, then the polymer chain $(A)_n$ will have a mass equal to nM_A. For a polymer containing M_i polymer chains of molar mass n_i, the number-average molar mass is:

$$\langle M_n \rangle = \frac{\sum_i n_i M_i}{\sum_i n_i} \qquad [2.1]$$

where $\sum_i n_i$ represents the total number of chains of the polymer.

The *mass-average molar mass* (or weight-average molecular weight) is calculated taking into account the mass w_i of the species that form the chains:

$$\langle M_W \rangle = \frac{\sum_i n_i M_i^2}{\sum_i n_i M_i} \quad [2.2]$$

Since the polymer is composed of a large number of chains of variable lengths and masses, its mass is described by a statistical function (such as Gaussian) that gives an indication of the distribution of molecular masses of molecules in the polymer. For this description, we use *dispersity*, which is defined as the ratio of the mass-average molar mass to the number-average molar mass, following the expression:

$$Đ(\text{or}\, Q) = \frac{M_W}{M_n} \quad [2.3]$$

A dispersity equal to unity means that the chains all have the same mass, that is, the distribution of the molecular masses would be reduced to a single mass corresponding to a single length of chain. In practice, the dispersity depends on the *number-average degree of polarization*, DP_n, which is the number of repeat units or monomers in the chain. It is defined as the ratio of the number-average molar mass of the polymer to the molecular mass m of the monomer:

$$DP_n = \frac{M_n}{m} \quad [2.4]$$

A polymer to be applied to achieve good performance must have the following qualities: a high degree of polymerization, low dispersity levels and a high molecular weight.

Depending on the structure of the molecules, a distinction can be drawn between polymers with linear molecules and polymers with branched and cross-linked molecules. When the molecules are linear, the chain is formed from a *backbone*, consisting of carbon atoms C or groups of atoms. On some carbon atoms, *side groups* are grafted, and the entire chain forms a linear structure with two endpoints, or chain ends. In polymers with complex

structures, the backbone of a chain has lateral chains of identical molecules, called *branches*. This structure is called a *branched structure*. Compared with the linear structure, the branched structure contains more than two chain ends. If the number of branches is large, then the polymer becomes more fragile in terms of its mechanics and less crystalline due to the short-distance disorder created by the branches. Another structure known as the *cross-linked structure* is obtained by connecting the molecules through chemical reactions between the atoms of the chains (chemical bridging) or through a strand of chains (chain strand bridging). The junction point is called the *cross-link point*. From a mechanical point of view, cross-linking strengthens the rigidity of the polymer and is sometimes voluntarily performed to avoid premature fracturing of the material when subjected to intense thermal or electrical stresses.

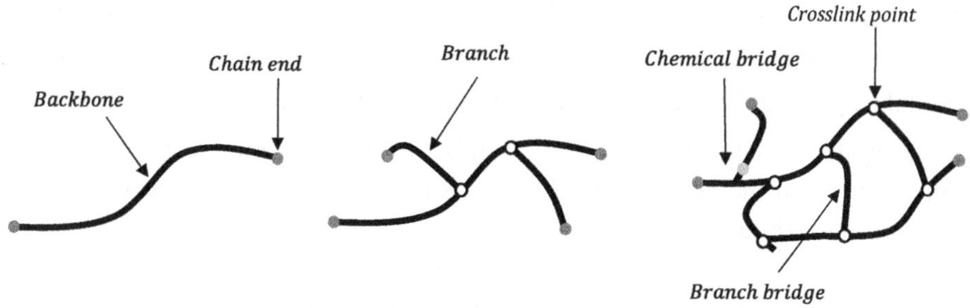

Figure 2.2. *Structures of polymers. For a color version of this figure, see www.iste.co.uk/nguyen/electronics1.zip*

One particularly interesting case in polymer structures is isomerism, in which two molecules having the same mass (called *isomers*) have different atomic arrangements. In the case of polymers with double bonds, the *cis–trans* isomer has two different configurations due to the geometry of the side groups.

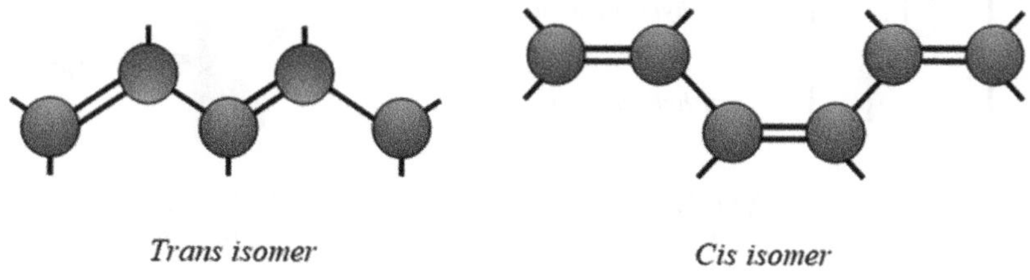

Figure 2.3. Trans *and* cis *configurations of polyacetylene*

The two configurations correspond to the different physical properties of the same polymer: the *cis* configuration provides for a flexible and amorphous polymer, while the *trans* configuration creates a mechanically hard and semi-crystalline polymer.

2.2.1. *Binding and hybridization of carbon*

In an organic material, carbon atoms form bonds with other atoms, including other carbon atoms. The formation of bonds between two atoms to form a molecule implies that the energy of the molecule obtained is less than the total energy of the individual atoms.

In its fundamental state, the electronic configuration of the carbon atom C with six electrons is $1s^2\,2s^2\,2p_x^1\,2p_y^1$. The orbitals $1s$ and $2s$ are either filled or occupied. In this configuration, two electrons are paired in the orbitals $2p_x$ and $2p_y$, and the third orbital $2p_z$ is unoccupied. Theoretically, each atom C can develop two bonds. In reality, the carbon atom is tetravalent, and the bonding mechanism is explained by one electron moving up from the orbital $2s$ to the orbital $2p$, as shown in Figure 2.4. The result of this transition is that four orbitals are occupied, each of them by one electron, which can then be combined with the remaining orbitals to form new combinations and create new bonds.

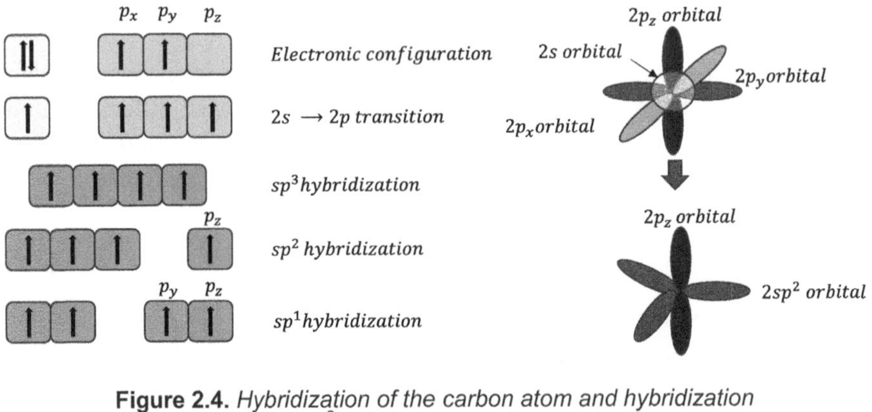

Figure 2.4. *Hybridization of the carbon atom and hybridization mechanism of sp^2. For a color version of this figure, see www.iste.co.uk/nguyen/electronics1.zip*

By linear combination of the four atomic orbitals $2s, 2p_x, 2p_y$ and $2p_z$, the carbon atom creates equivalent hybrid orbitals sp^3, each containing one electron. The orbitals obtained have axes, creating the same angle of 109° 28' between them, oriented along the axes of a tetrahedron whose center is occupied by the carbon atom. The polymers formed by the sp^3 hybridization of carbon have simple bonds $-C-C-$ in the backbone. These bonds are called σ *bonds* and are formed by the overlapping of the orbitals along the same axis. Such polymers are not SCs because the electrons of the molecules are localized.

The hybridization of the orbitals $2s, 2p_x$ and $2p_y$ leads to three equivalent coplanar hybrid atomic orbitals sp^2, whose axes form an angle of 120° between them. The atomic orbital $2p_z$ is unchanged and its axis remains perpendicular to the plane. To form a bond between two carbon atoms, two hybrid orbitals sp^2 of these two atoms are combined, with each orbital providing an electron. The other two orbitals form bonds with the other atoms (such as hydrogen). The hybrid bonds of the carbon atoms are along the plane of the molecule and form a σ bond (or overlapping of orbitals of the same axis). This bond is rigid and forms the backbone of the molecule. On the other hand, the axes of the two orbitals $2p_z$ of the atoms, each containing an unpaired electron, are parallel and may overlap, leading to the formation of a *bonding molecular orbital* π and an *anti-bonding orbital* π^*. These two orbitals form a π bond (see Figure 2.5).

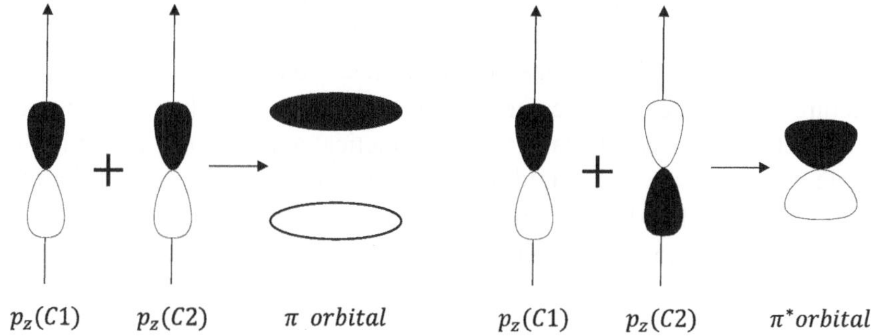

Figure 2.5. *Atomic orbitals π and π^* of the π bond*

The polymers formed by the sp^2 hybridization of carbon atoms have double bonds $-C=C-$ in the backbone. The electrons of the bonding orbital are called π electrons and are *delocalized*; that is, they no longer remain bound to the electrons from which they originated. The *delocalization* of electrons is not limited to one bond but may involve several bonds created between carbon atoms. The overlapping of the orbitals $2p_z$ of two adjacent atoms implies that an electron associated with one of the atoms no longer belongs specifically to that atom and that the free electron of the π bond is present in the geometric space occupied by the two atoms. This phenomenon is called *conjugation*, a particular characteristic of covalent bonds.

The sp^1 hybridization of carbon atoms leads to triple bonds $-C\equiv C-$ in the backbone

2.3. Conjugated polymers

A *conjugated system* consists of atoms bound by covalent bonds with at least one delocalized π bond.

There are several types of conjugations. Conjugations of type $\pi-\sigma-\pi$ are typical of polymers and organic materials with electronic properties comparable to inorganic SCs. Such a system consists of single and double or triple bonds. The delocalization of π electrons occurs at the level of the double bonds of carbon atoms. The overlap of these orbitals extends this delocalization to the entire backbone and the conjugation is theoretically carried out over the entire polymer chain. The simple bond between two atoms is a σ bond and has no delocalized electrons. The alternating structure is nevertheless necessary to obtain the proper mechanical properties of the material needed to shape films in practical applications. In reality, the double bonds $C=C$ are rigid because they prevent the molecule from rotating. Simple bonds *C–C* give the chain structure more flexibility and allow the material to be more easily shaped and used in devices.

It is important to note that the delocalization of electrons is not unlimited, otherwise polymers would behave like metals. Since the chains are not perfectly linear, their length is irregular and they contain defects.

In reality, the delocalization of electrons is limited within the material to domains of varying sizes, from roughly ten to a few hundred double bonds, called *conjugation lengths*. The electrons are free and move along the chains over an average distance equal to the conjugation length. This parameter is important and influences the electrical and optical properties of the organic material.

We will examine two examples of conjugated polymers: polyacetylene and benzene.

2.3.1. *Polyacetylene*

Acetylene, whose chemical formula is $H–C \equiv C–H$, is composed of two carbon atoms bound by a triple bond. Each carbon atom C is also bound to a hydrogen atom H.

Examined in greater detail, we find that in the molecule, there are:

– two σ bonds of type $sp–s$ for the $C–H$ bonds;

– one σ bond of type $sp–sp$ for the $C–C$ bond;

– two π bonds of types $p–p_y$ and $p–p_z$.

Polyacetylene has two molecular structures with two identical energy levels: *cis* and *trans*.

Figure 2.6 shows the structure of polyacetylene in the *trans* configuration.

We will consider the orbitals p_z, where overlapping forms the π bonds that allow for the delocalization of electrons along the entire polymer chain length. The π band contains two bands: bonding π and anti-bonding π^*. Each band contains two electrons per atom: the π band is filled and the π^* band is empty.

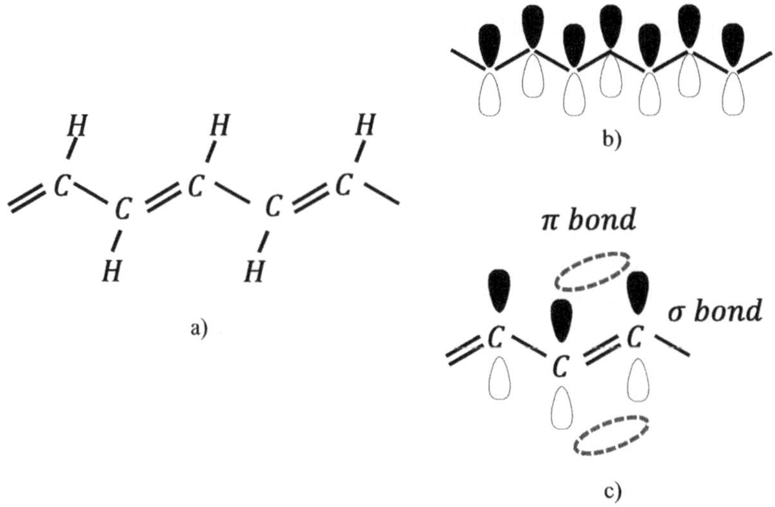

Figure 2.6. *Structure of polyacetylene: a) chemical bonds between atoms; b) orbitals of carbon atoms; c) π and σ bonds of the chain. For a color version of this figure, see www.iste.co.uk/nguyen/electronics1.zip*

Theoretically, polyacetylene is conductive because, in this configuration, electrons are free to move along the chain. In reality, this polymer is an SC because of the existence of an energy gap, which is a potential barrier that electrons must cross.

This gap is similar to the forbidden band of traditional SCs, whose origin was explained by Heeger *et al.* (1988). According to the authors, the delocalization of electrons does not extend to the entire chain because the lengths of single and double bonds are not identical. A limit to the electron delocalization must necessarily exist. On the other hand, a one-dimensional and regular molecular chain (such as a chain of dihydrogen molecules) does not exist because the adjacent molecules cause interactions that modify their energy.

This change results in the instability of the chain, which deforms to maintain stability. This process is called *Peierls distortion*, which in turn introduces a new energy band that forms the gap of the chain. Therefore, a conjugated polymer cannot be conductive in the way metals are but instead behaves like an SC.

2.3.2. Benzene

The benzene molecule contains six carbon atoms arranged on the vertices of a regular hexagon (see Figure 2.7). Each atom is bonded to a hydrogen atom and to the two neighboring carbon atoms C by a single bond and a double bond.

The bonds are of σ-type between the atoms of the molecule and of π-type in the double bonds between carbon atoms. The orbitals $2p_z$ of the atoms C overlap two over two, forming three π molecular orbitals each and three empty π^* molecular orbitals. The π electrons are delocalized across the entire molecule, forming two symmetrical toric clouds relative to the molecular plane.

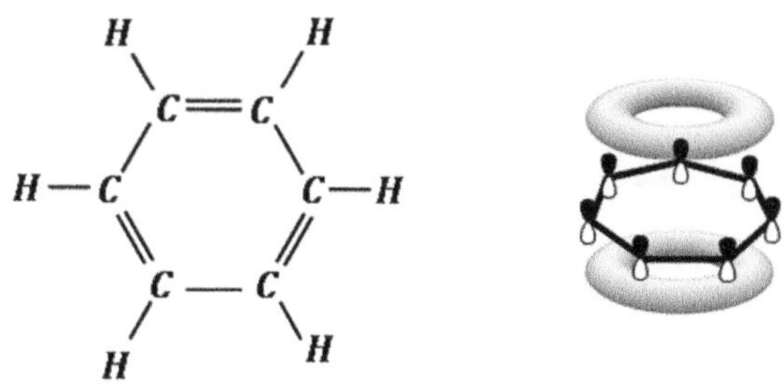

(a) (b)

Figure 2.7. *Structure of benzene: a) chemical bonds between atoms; b) orbitals of carbon atoms and π electrons of the molecule*

Like polyacetylene, aromatic polymers (such as poly(phenylene)) also have two molecular structures: *aromatic* and *quinonic* (see Figure 2.8). However, the energy states of these structures are not equivalent.

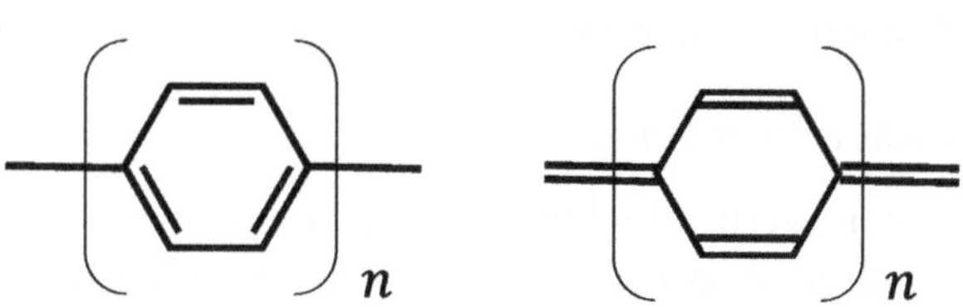

Aromatic *Quinoid*

Figure 2.8. *Molecular structures of poly(phenylene)*

2.3.3. *Deposition of polymer films*

The polymers used for electronic components are obtained in powder form and are soluble in the proper solvents. The polymer solution is deposited on thin films on glass or plastic substrates (usually poly(ethylene terephthalate) or PET) through several techniques. Figure 2.9 illustrates the principle of these techniques.

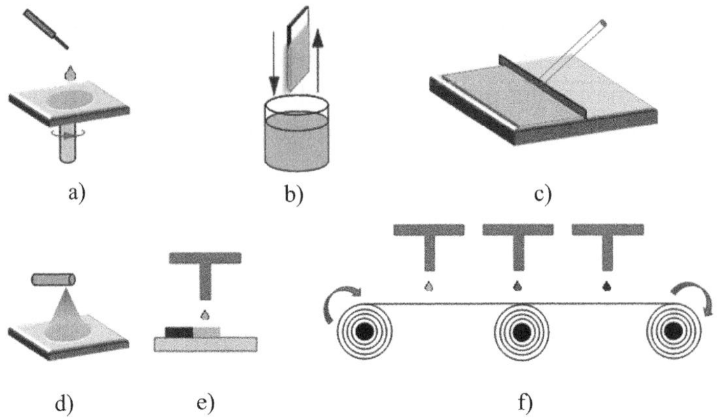

Figure 2.9. *Different techniques for deposition of polymer films in a solution: a) by spin-coating; b) by dipping; c) by doctor-blading; d) by spray-coating; e) by inkjet-printing; f) by roll-to-roll printing. For a color version of this figure, see www.iste.co.uk/nguyen/electronics1.zip*

2.3.3.1. *Deposition by spin-coating*

The polymer solution is deposited on a substrate that spins at high speed. The liquid spreads over the substrate and forms a thin film after drying. The thickness of the film partly depends on the speed of rotation, which can be calibrated to obtain the desired value.

2.3.3.2. *Deposition by dipping*

The substrate is vertically immersed in a polymer solution and then removed at a controlled speed. The film is formed after drying and covers both sides of the substrate.

2.3.3.3. *Deposition by doctor-blading*

The polymer solution is deposited on the surface of the substrate and then spread evenly using a doctor blade that moves at a controlled speed. The thickness of the film can be adjusted by the position of the blade relative to the surface of the substrate.

2.3.3.4. *Deposition by spray-coating*

The polymer solution is vaporized and then sprayed via an atomizer onto the substrate, forming a film. The homogeneity and morphology of the film can be controlled by the concentration of the solution, size of the droplets and speed of the fluid projection on the substrate.

2.3.3.5. *Deposition by inkjet printing*

The polymer solution is used like ink in a printer and supplied through the printer head. This technique makes it possible to print films of various shapes and sizes and to manufacture components whose characteristics are devised using software for writing or drawing.

2.3.3.6. *Deposition by roll-to-roll printing*

This technique allows the film to be deposited on a plastic substrate wound around a roll, which is then unwound by the rotation of a second roll. The deposition of the films can be done by inkjet printer heads. The unrolling of the substrate done at high speeds makes it possible to print the films in large quantities and mass produce the components at a low price.

2.4. Energy bands

The concept of energy bands in organic materials is quite similar to the concept of bands in inorganic SCs. With this in mind, we can consider the energy of a system consisting of two hydrogen atoms, A and B. When two atoms with the same energy are brought into contact, the energy of the molecule can be determined by the energy of the isolated atoms, according to the theory of linear combination of atomic orbitals. In other words, the wave function associated with an electron in the system $(A+B)$ can be approximated by a linear combination of the wave functions of atoms A and B. If the functions associated with the atoms are $\Psi_{1s}(A)$ and $\Psi_{1s}(B)$, respectively, we can write:

$$\Psi_{\pm} = \Psi_{1s}(A) \pm \Psi_{1s}(B)$$

Thus, there are two molecular orbitals:

– Bonding orbital: $\Psi_{+} = \Psi_{1s}(A) + \Psi_{1s}(B)$;

– Anti-bonding orbital: $\Psi_{-} = \Psi_{1s}(A) - \Psi_{1s}(B)$.

The amplitude of these orbitals is shown in Figure 2.10.

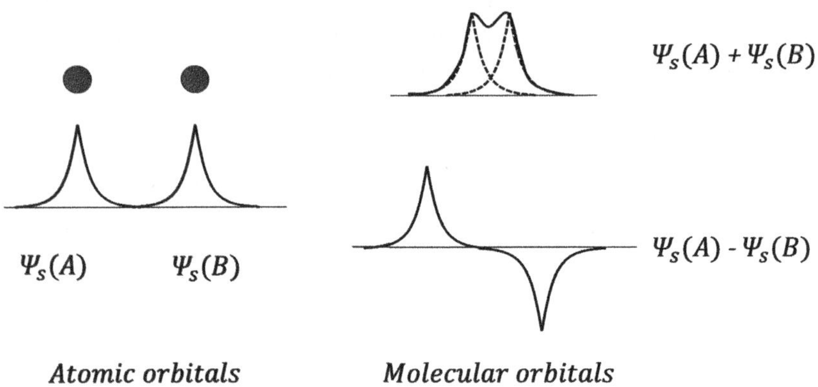

Figure 2.10. *Wave functions of atomic and molecular orbitals*

In this scheme, the two electrons occupy the bonding orbital of the molecule, which has a lower energy than that of the anti-bonding orbital. In the organic solid composed of a large number of molecules, the superposition of the molecular orbitals leads first to an increase in orbital number and then to an increase in energy bands when the energy of these orbitals becomes indistinguishable.

To study the transport of free charge carriers, we must take into account the conjugation length in the material. Free charges only travel a limited distance, within which these electrons are delocalized. This length varies depending on the materials. It is of the order of a few dozen constituent units. However, when the number of conjugated molecules increases, the energy gap between the orbitals π and π^* decreases. Extending beyond the conjugation length, this gap becomes stabilized.

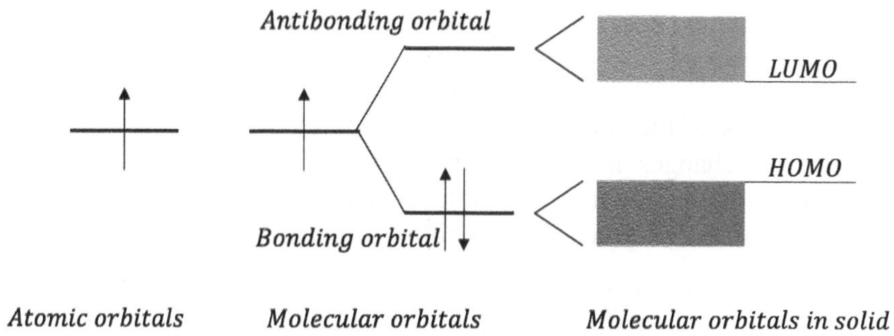

Figure 2.11. *Energy levels of the isolated atom, the molecule formed by two atoms, and an organic solid. For a color version of this figure, see www.iste.co.uk/nguyen/electronics1.zip*

The highest occupied molecular orbital of the VB is called the *HOMO* (Highest Occupied Molecular Orbital), and the lowest unoccupied molecular orbital of the CB is called the *LUMO* (Lowest Unoccupied Molecular Orbital) (see Figure 2.11). These two levels correspond to the top level E_V of the valence band (VB) and the bottom level E_C of the conduction band (CB) in the scheme of energy bands of inorganic SCs. The difference in energy between the LUMO and HOMO levels is the gap E_G of the organic material. For most organic materials used in electronic applications, the gap value varies between 2 and 3 eV, which means that organic materials are classified as SCs from an electrical point of view. Thus, it is possible to refer to them as organic SCs.

Having thus established the energy band model, the physical processes in organic materials can be analyzed and explained based on the findings from the study of inorganic SCs. The same physical parameters already defined for these materials will be used to describe the characteristics of organic SCs, as well as those of the devices that use them as active materials.

2.4.1. *Concepts of solitons and polarons*

In a conjugated system such as polyacetylene, the permutation of single and double bonds when transforming the *cis–trans* configurations occurs in two cases:

– the permutation requires an expenditure of energy to occur. Such a system is known as *non-degenerate*;

– the permutation does not change the energy of the states. Such a system is known as *degenerate*.

On the basis of the concept of minimum energy, the conjugated system adapts to the changes in its electronic structure by conserving its initial energy or maintaining it at its lowest level. This is made possible by a local deformation of the polymer chain, leading to the formation of solitons and polarons.

In a degenerate conjugated polymer, breaking of the conjugation can occur at double bonds when two unpaired electrons are created. Each charge is actually a charged radical ion. Schematically, depending on the position of the double bond, there are two molecular structures that correspond to two energy states or *phases* of equal energy denoted by A and B. A perturbation to the system causes it to switch from one phase to another. A *soliton* is defined as a defect or an excitation between phases A and B, which can cause a change in phase during its movement. It is a charged radical ion. A soliton is associated with an energy state, located in the center of the gap. The level can be occupied by zero, one or two electrons. Without an electron, the soliton is positively charged and has no spin. With one electron, the soliton is not charged and has a spin equal to $\pm\frac{1}{2}$. With two electrons, the soliton is negatively charged and has no spin. The states of the soliton are shown schematically in Figure 2.12.

A polaron can be defined as the association between a charged soliton and a neutral soliton. This association implies that the polaron has a charge and a spin. Since each soliton introduces an energy state, the polaron has two energy states that may or may not be occupied by electrons. This property gives rise to another definition of the polaron, which can be formulated as the association of an electron with a distortion via a polarization. This definition on the basis of solitons allows us to distinguish two types of polarons: positively charged polarons (of charge $+|e|$ and spin $+\frac{1}{2}$) and negatively charged polarons (of charge $-|e|$ and spin $+\frac{1}{2}$).

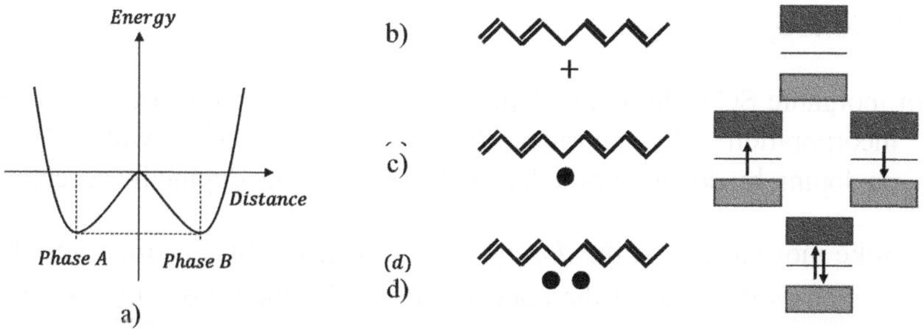

Figure 2.12. *Representation of solitons: a) energy of a dimer consisting of two states A and B without excitation; b) neutral soliton; c) soliton charged with an electron; d) soliton charged with two electrons. For a color version of this figure, see www.iste.co.uk/nguyen/electronics1.zip*

Similarly, a *bipolaron* is an association of the two charged solitons and introduces two energy states into the gap in the organic material. A bipolaron can be positively charged when it combines two positive solitons (of charge $+ 2\ |e|$ and with no spin). It is negatively charged when it combines two negative solitons (of a charge of $- 2\ |e|$ and with no spin). The schematics for the formation of polarons and bipolarons are shown in Figure 2.13.

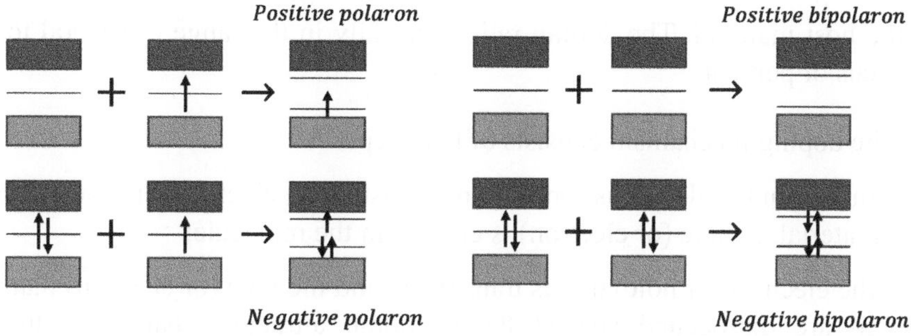

Figure 2.13. *Formation of bipolarons by solitons. For a color version of this figure, see www.iste.co.uk/nguyen/electronics1.zip*

We can note that the creation of solitons and polarons is related to the non-degenerate nature of the conjugated system, which is due to the changes of the bond lengths (as with polyacetylene). Therefore, in conjugated systems with benzene rings, this characteristic is absent and there will be no solitons or polarons.

2.4.2. Concept of doping

In inorganic SC technology, doping seeks to increase conductivity through the incorporation of impurities that supply the CB with electrons (N-type doping by donors) or the VB with holes (P-type doping by acceptors).

Unlike inorganic SCs, the *doping ratio* of organic SCs is often very high. It is defined as the ratio of the concentration of doping molecules n_D to the concentration of molecules n_H of the host material. In molar concentrations, the expression of the molar ratio is written as:

$$mol\% = 100 \times \frac{n_D}{n_D + n_H} \qquad [2.5]$$

In molar mass concentrations, it is written as:

$$wt\% = 100 \times \frac{m_D}{m_D + m_H} \qquad [2.6]$$

where m_D is the mass of dopant molecules and m_H is the mass of molecules of the host material. The doping ratio is usually in the range of several to a few tens of percent.

The doping mechanism consists of two steps:

– the dopant molecule is ionized and transfers an electron (or hole) to the host material. A hole (or electron) is created in the molecule;

– the electron (or hole) that is transferred and the hole (or electron) that is created are dissociated against the Coulomb attraction that binds them together. The charges released contribute to increasing the conductivity of the doped material.

For organic SCs, the principle is identical and doping makes it possible to not only improve the conductivity of the material but also to facilitate the charge injection into the devices through the transport layers.

For an organic SC, N-type doping consists of transferring electrons of the dopant to the LUMO of the host and P-type doping consists of extracting electrons from the HOMO of the host to the dopant, generating holes in this band.

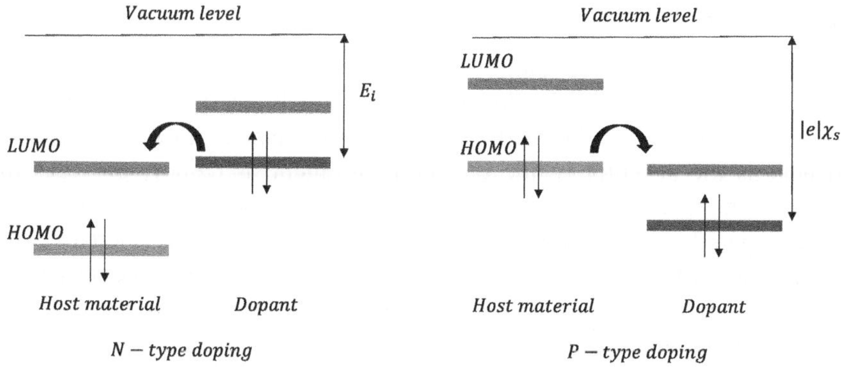

Figure 2.14. *Doping mechanisms for N-type and P-type organic SCs. For a color version of this figure, see www.iste.co.uk/nguyen/electronics1.zip*

For the doping of an organic SC of a defined conduction type (*N* or *P*), it is essential to choose the doping material with the appropriate energy characteristics. Let E_i be the ionization energy and $|e|\chi_S$ be the electron affinity of an SC. For N-type doping, the dopant molecule must have low ionization energy to be able to transfer electrons to the LUMO level of the host material. For P-type doping, it must have a high electron affinity to be able to receive electrons from the HOMO level of the host material. The scheme of the mechanisms of charge transfer by doping is shown in Figure 2.14.

However, the structures of organic and inorganic SCs are different, and therefore the properties of the doped materials are also different. The most important difference between the two types of SCs lies in the electrical conduction mechanism of free charge carriers. Indeed, in inorganic SCs, free electrons and holes move in their allowed bands and doping increases the concentration of these carriers, and thus their conductivity. In organic SCs, the charge transport occurs via the hopping process between the localized states of the energy distributions in the gap of the SC. In this case, the doping is intended to increase the number of charges that can hop. In other words, doping allows polarons and bipolarons to be formed and creates energy bands, which allows an easier charge movement when an electric field is applied to the material. As a result, the electrical conductivity of organic SCs increases and can reach values comparable to those of metals.

2.4.2.1. *Chemical doping*

The doping is carried out by a *redox (reduction-oxidation) reaction*, between the polymer and the dopant, during which electrons are transferred between the materials. The oxidizing agent or oxidant captures the electrons and acts as the acceptor, while the reducing agent or reducer provides the electrons and acts as the donor. The interaction between the chains of the conjugated polymer and the donor or acceptor atoms increases the concentration of the π electrons and, consequently, the conductivity of the polymer. The electrical neutrality of the doped polymer is preserved by the creation of counter-ions (anions or cations) in the polymer matrix.

For P-type doping, which is the most common type, the oxidants used are usually halogen gases, such as iodine or bromine. In the case of polyacetylene, the first doping was done using iodine, which is an oxidizing dopant. The oxidation reaction is written as:

$$[CH]_n + \left(\frac{3x}{2}\right) I_2 \rightarrow [CH]_n^{x+} + x\, I_3^- \qquad [2.7]$$

The charges from the anions I_3^- are delocalized along the polyacetylene chains and contribute to increasing the electrical conductivity of the polymer. In order to obtain high conductivity using this doping method, it is necessary to use a large amount of dopant, and the usual doping ratio will vary between 10% and 30% – very high in comparison with the concentration of impurities in inorganic SCs.

For N-type doping, alkali metals such as potassium and lithium and the compounds of these metals are used. These atoms are easily diffused within the polymer matrix, making the material unstable. Another possibility for N-type doping is to use molecular compounds with HOMO levels higher than the LUMO level of the SC, to allow electrons to be transferred directly. However, such compounds are also unstable for oxidation.

Chemical doping is a reversible process and is carried out in an aqueous medium. For small molecules, the doping is performed under vacuum through the co-evaporation of the organic material and the dopant. The doped polymer is often unstable and degrades upon contact with air. This instability is due to the small size of the atoms or molecules of the dopant that tend to diffuse inside the polymer. To improve the stability of doped polymers, large molecules used as donors or acceptors (molecular doping) can improve their conductivity by several orders of magnitude.

2.4.2.2. *Electrochemical doping*

Electrochemical doping is carried out in electrolytic solutions into which two electrodes are immerged. One of the electrodes is coated with a polymer film that will be doped. This polymer must be insoluble in the electrolyte. This electrode constitutes the anode for a chosen N-type doping (or the cathode for a chosen P-type doping). The second electrode (counter electrode) allows voltage to be applied. The electrons (or holes) are injected into the polymer under the current flow, and at the same time, the ions of the electrolyte are also diffused within the film to compensate for the charges.

Compared with chemical doping, electrochemical doping has the advantage that the doping ratio can be controlled by controlling the reaction rate and the applied voltage.

There are other doping processes for obtaining *extrinsic polymers* with high electrical conductivity through the incorporation of nanoparticles into an insulating polymer matrix. We will study these materials as part of the category of nanocomposites.

2.5. Small molecules

In a polymer formed from constituent units or monomers, the bonds between these units are covalent, which means that they are strong bonds. In small molecules, the bonds between the molecules are of the van der Waals type, meaning they are weak bonds. As a result, the interactions between molecular orbitals are weak, and the energy bands of the molecular solid (HOMO and LUMO) are narrow, formed from discrete energy levels and not in continuous bands, as with polymers. Therefore, the movement of the free carriers in the bands requires an energy supply to enable them to reach the different energy levels in the bands. The carrier mobility in small molecules is generally higher than that of carriers in polymers.

Organic molecules can bind together to form *oligomers* with a limited number of repeat units. As with polymers, when the number of their constituent units increases, the energy gap between LUMO and HOMO levels decreases, which leads to changes in the optical and electrical characteristics. In addition, the stability of molecules decreases when the size of molecules becomes large. One example of such systems can be found in the small molecules formed from benzene cycles, which may include two (naphthalene), three (anthracene), four (tetracene) or five cycles (pentacene) (see Figure 2.15). The last two of these materials in particular are used in organic field-effect transistors. Other combinations of small molecules by alternating oligothiophene acceptor and donor $A - D - A$ molecular groups made up of four to nine units make it possible to modulate the absorbance and the LUMO and HOMO energy levels, and to use them as absorbers in solar cells.

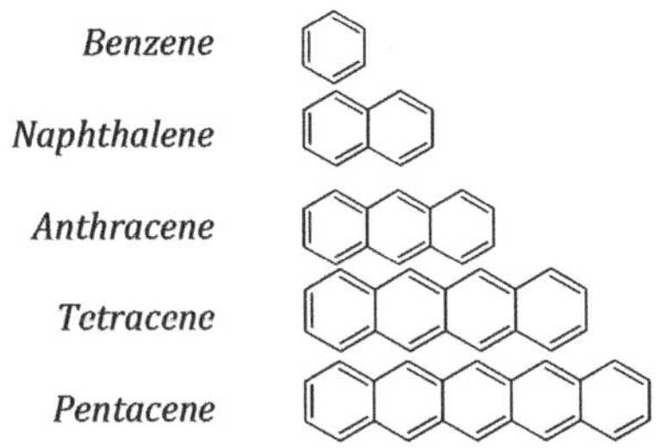

Figure 2.15. *Molecular structure of small molecules of benzene cycles*

Compared to polymers, small-molecule films have better defined molecular structures and contain fewer defects due to the absence of chain ends. On the other hand, although the deposition of the films is performed by evaporation of powders under vacuum and not by solutions as with polymers, the quality of the films is generally good and easy to control, without significant variations in properties as with the case of polymers. Finally, the charge carrier mobility is higher than that measured in polymers due to better crystallinity.

Figure 2.16 gives a preview of the polymers and small molecules used in organic electronic devices. There are numerous materials that are synthesized and studied, but few of them are actually suitable for practical applications due to their complicated elaboration, cost or performance.

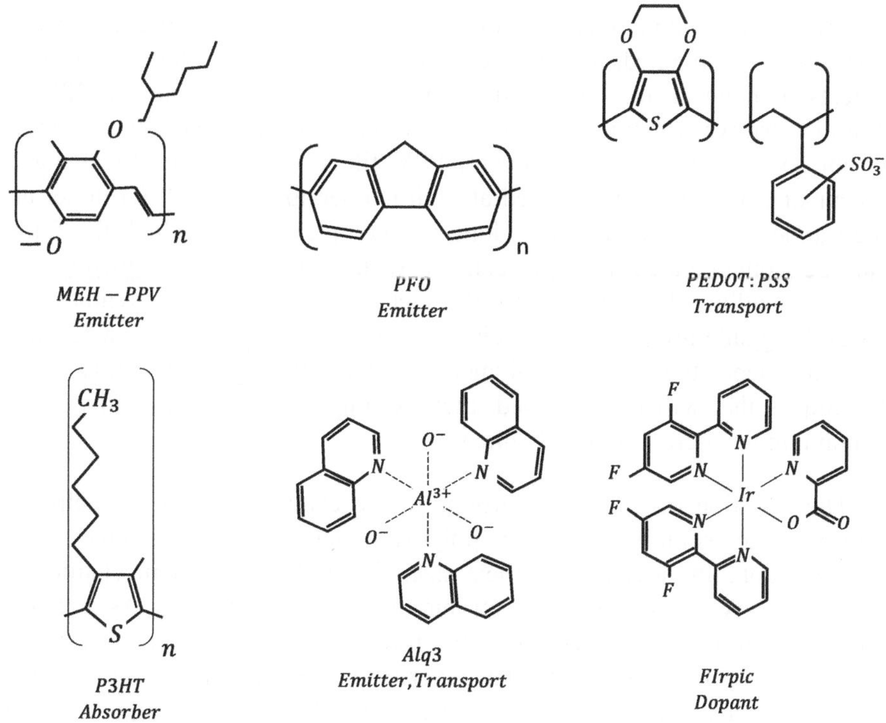

Figure 2.16. *Examples of polymers and small molecules and their use in electronic devices*

2.6. Design and engineering of organic materials

To obtain organic materials with good mechanical and electrical properties, a number of criteria must be fulfilled. First of all, the σ bonds make the skeleton rigid, which affects the flexibility of the organic film. Additionally, in order to achieve a reasonable delocalization of the π electrons, it is necessary to have an acceptable conjugation length. Most of the organic materials used in the devices are conjugated systems with alternating saturated and unsaturated bonds to ensure good mechanical strength and also a good electrical conductivity of the materials.

In practice, organic SCs are used as thin films and film deposition using solutions remains a simple and economical method. To facilitate and improve the solubility of materials in solvents, the alkyl groups are grafted onto the skeletons of the polymer chains. Indeed, a good dissolution of the material in a suitable solvent allows for uniform films with high-quality surfaces to be obtained, that will form interfaces with other layers with few defects or none at all. For example, poly(phenylene vinylene) or PPV, a polymer that was often used in the study of early organic electronic devices, is not soluble directly in a solvent. PPV films are deposited by using a precursor solution, then annealed at high temperatures (~ 300°C) under an inert atmosphere. The process is long and complex, during which the films can be contaminated through contact with different environments and through handling. Derivatives of PPV, such as MEH-PPV, were synthesized by grafting side groups onto the chain to dissolve the polymer powder in a solvent. Deposition, using polymer solutions through commonly used techniques that will be described later, is simple and allows quality films containing few defects to be obtained.

The insertion of the lateral groups to the chain, often alkyls, helps to improve the solubility of the polymers but can also change their conformation. In fact, due to their size, the molecules can mechanically deform chains and reduce the electronic overlap of the orbitals. As a result, the conjugation is reduced and the transport of charges decreases in comparison with a polymer that has a perfectly flat skeleton. To remedy this deformation, using heteroatoms in side groups reduces the chain deformation and increases the conductivity of the polymer. As an example, consider poly(hexylthiophene) or P3HT, which is a popular polymer often used as an active layer in devices. Alkyl side chains allow the polymer to be dissolved in many common solvents. However, the organization between chains leads to several structures whose physical characteristics are very different from each other. We can distinguish three types of chain formation or dyads, according to the arrangement of molecules and side groups: head-to-tail (HT), head-to-head (HH) and tail-to-tail (TT). The combinations of dyads in four triads are shown in Figure 2.17. P3HT organized as type HT-HT is known as *regioregular* when the chains are stacked regularly, forming a lamellar structure with a minimum footprint of alkyl chains. The polymer chain is flat overall and efficiently transports charges with high mobility. $\left(\sim 0.1\ \mathrm{cm^2V^{-1}s^{-1}}\right)$ Other forms of organization lead to non-regioregular P3HTs, which have a more disorderly structure, and less effective charge transport with low mobility of carriers $\left(\sim 0.1\times 10^{-3}\ \mathrm{cm^2V^{-1}s^{-1}}\right)$. As a general rule, the solubility of materials improves all the more when the alkyl chain is longer. Nevertheless, problems related to the steric effect of the side chains, as well as changes in morphology, can significantly affect the optical and electrical transport processes in the devices, and therefore will have to be considered. The photophysical properties of the polymer material strongly depend on the chain arrangement in the structure, which itself depends on the method of preparation of the chemical and the shaping of the thin films. For example, intra-chain interactions between constituent units lead to type J molecular aggregate behavior, while inter-chain Coulomb interactions (interactions between chains) lead to type H- aggregate behavior. For the example of P3HT, a dyad of type HT corresponds to a type J chain aggregation, while a dyad of type HH-HH or T-T corresponds to a type H chain aggregation.

Another important point to consider for designing polymers is the molecular weight. In general, a polymer with high molecular weight has a good conjugation length, typically from 8 to 10 constituent units, and good solubility in commonly used solvents. By contrast, the solubility of a polymer with low molecular weight is poor, and precipitates are formed as it is polymerized. In addition, the conjugation length is short, with the formation of numerous chain ends that lead to creating traps for the charge carriers. These defects influence transport and optoelectronic processes in organic components.

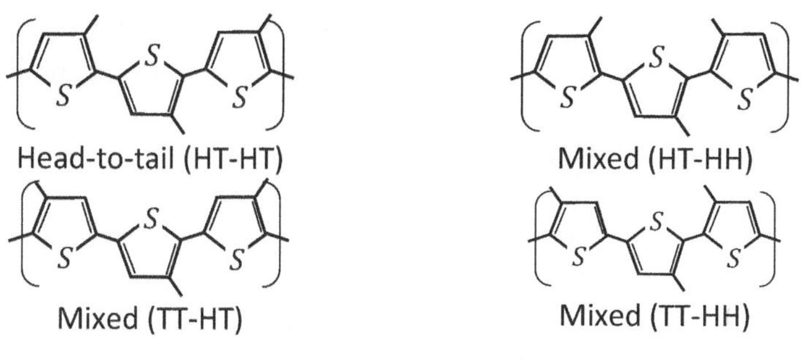

Figure 2.17. *Types of chain arrangement in poly(hexylthiophene)*

2.7. Hybrid materials or nanocomposites

A composite material is defined as an assembly of two or more different materials and has remarkable physical properties not exhibited by the individual components. We can distinguish two methods for forming composites depending on the nature of the materials used. In the first case, a constituent material is used as a physical medium on or in which the other material is deposited, and the properties of the latter will be exploited in the use of the composite. In the second case, the composite material has the combined properties of the constituent materials, and is used to achieve better qualities and performance than those of the individual components. The term "nanocomposite" is used to refer to composite materials consisting of components whose size is at the nanoscale.

Inorganic SCs are robust and stable. Their chemical and electronic structure is not significantly affected by contact with the external environment. Oxidation reactions may potentially occur when an SC contacts the ambient atmosphere, forming an oxide layer on the surface of the material. Nevertheless, the physical properties of the material do not fundamentally change, and contamination can be eliminated by adequate surface treatment. In contrast, organic SCs are fragile and unstable. Their electronic structure can be greatly modified through contact with the outside environment or when the temperature is high. The bonds between the atoms or molecules can be broken or replaced, thus modifying the conjugation of the polymer chains.

These changes will lead to the loss of the physical properties of the material, and as a result, a decrease in the performance of the devices that use it. Table 2.1 provides a schematic summary of the differences between inorganic and organic SCs in how these materials are used and applied.

Organic SCs	Inorganic SCs
Flexibility	Rigidity
Lightness	Heaviness
Wide surface	Small surface
Low cost	High cost
Instability	Stability

Table 2.1. *Comparison of the properties of organic and inorganic SCs*

Hybrid materials composed of inorganic and organic SCs are designed to compensate for the weak points of organic materials, and also to obtain some original and remarkable physical properties. Two types of nanocomposite materials can be distinguished: polymer matrix composites and nanomaterial composites.

2.7.1. *Polymer matrix nanocomposites*

In these materials, a polymer deposited on a thin film is used as a matrix to ensure the composite is cohesive after inorganic nanoparticles are added. Several polymers, both conjugated and non-conjugated, can be used as a matrix, depending on how the composite is intended to be used. The inorganic particles are generally SCs or compounds with particular optical or electrical properties, such as titanium dioxide (TiO_2), zinc oxide (ZnO) or silicon dioxide (SiO_2). The incorporation of these particles makes it possible to improve or modify the properties of the polymer, making it more efficient. More recently, new nanomaterials such as carbon nanotubes (CNTs), graphene and metal nanoparticles have been used in these composites to create further remarkable functions or properties. One of the difficulties in synthesizing polymer matrix composites is the inhomogeneous dispersion of nanoparticles in the matrix. In fact, at the nanoscale, particles tend to aggregate during their preparation, forming aggregates in the matrix. As a result, controlling the physical properties of the material becomes problematic. One possible solution to the problem is the use of surfactants to improve the dispersion and distribution of the particles.

2.7.2. *Nanocomposites with nanomaterials*

Hybrid nanomaterials are *colloidal nanocrystals* or *quantum dots* (QD). They are chemically synthesized from SC nanoparticles. These nanocrystals are formed with an inorganic core and a layer of organic *ligands* (*shells* or *crowns*) adsorbed to the surface by a chemical reaction between the two parts. The optical or electrical properties of the composite are due to the SC in the core, while the shell acts as a passivation layer and, in some cases, provides specific functions to the core. The characteristic of these composites is their *quantum confinement* effect, allowing their optical properties to be paired with their size and composition. In fact, quantum confinement tends to increase the energy of the confined materials in relation to their equivalent mass. That is, when the particle size decreases, the SC energy gap increases and the energy of the photons emitted by the small particles becomes higher. By varying the size of the crystals, it is possible to change the emission spectrum of the material and thus obtain the emission of several different colors using the same emitting material.

2.7.3. *Preparation of nanocomposites*

Several techniques for the preparation of polymer-based nanocomposites were used to obtain materials suitable for the realization of organic electronic components. The problem of particle dispersion remains fundamental to ensuring the quality of the composite material. Indeed, it is essential to have a homogeneous distribution of the particles in the polymer matrix, because if these particles are organized into aggregates, the unique properties that are expected will be significantly modified. The primary processes are nanoparticle dispersion, surface modification, covalent bonding and the formation of double layers.

2.7.3.1. *Dispersion method*

Nanoparticles are incorporated into the polymer that has been already prepared in a solution with an appropriate solvent. This is then mixed thoroughly, and by deposition of the solution on a substrate, the composite is obtained in the form of a film after the evaporation of the solvent (see Figure 2.18). To obtain a homogeneous dispersion of the nanoparticles in the solvent, the solution is constantly stirred by magnet or ultrasonication. Generally, the homogeneity of the composite material is poor, because the nanometric size of the particles easily allows them to agglomerate by forming aggregates in the polymer matrix and changes the morphology of the film obtained. To remedy this, the use of surfactants helps to reduce the aggregation effect and to obtain a better dispersion of particles and films of a more homogeneous structure and morphology.

This method is commonly used to realize the active layers in organic light-emitting diodes (OLEDs), organic solar cells (OSCs) and organic photovoltaics (OPVs). Among the oxide nanoparticles, $Ti0_2$ is particularly interesting as it is an electron transporter and improves the electronic conduction of the polymer with which it forms the composite material. Light-emitting diodes using an active composite layer made with MEH- PPV and nanocrystals of $Ti0_2$ have shown a higher external quantum efficiency than that of diodes with active layers of MEH- PPV alone. Other oxides, such as Zn0 and $Si0_2$, have also been used in composites with conjugated polymers, such as poly(hexylthiophene) (P3HT), poly(vinycarbazole) (PVK) and poly(fluorene) (PF), which possess good optical and electronic properties. The use of carbon nanomaterials such as graphene (and its derivatives, such as graphene oxide (GO) and reduced graphene oxide (rGO)) or carbon nanotubes as fillers in composites has been widely studied. These materials have excellent mechanical and electrical properties, and, when incorporated into the polymer, they contribute to strengthening the structure and stability of the composite. At the same time, the charge carrier transport in the material is improved, allowing the performance of devices to be significantly increased.

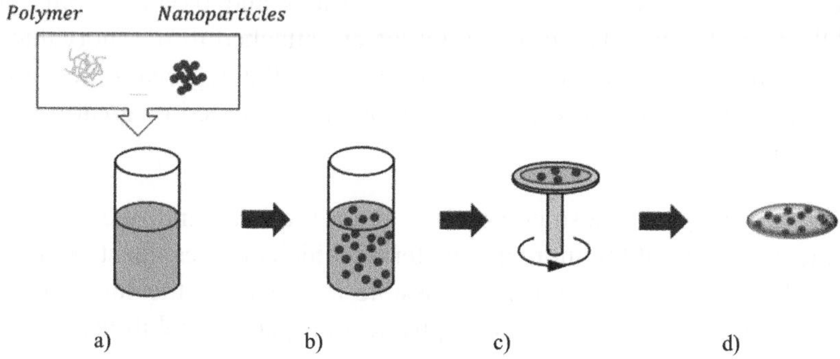

Figure 2.18. *Preparation of composites by dispersion method: a) materials; b) dispersion of nanoparticles; c) deposition by spin-coating; d) composite film. For a color version of this figure, see www.iste.co.uk/nguyen/electronics1.zip*

2.7.3.2. *Surface-modified method*

This technique involves modifying the surface of the nanoparticles to become more compatible with the polymer of the matrix. First, a polymerization initiator is grafted onto the surface of the particles and then polymer chains are grown from the initiator by polymerizing the monomers. In this way, the polymer molecules are attached to the surface of the particles by chains of the same polymer as the polymer of the matrix. This method allows for the control of the grafting density of the particles, which is usually high. Another grafting technique involves bringing the polymer chains that are already synthesized into contact with the surface of the particles. The reaction between the chain ends and the surface makes it possible to attach the polymer molecules to the nanoparticles. Due to the steric effect, the grafting density on the particles is lower than that obtained by polymerization *in situ*.

2.7.3.3. *Covalent bonded route method*

The technique used to link the organic molecules directly to the inorganic material involves the construction of covalent bonds between the two materials. This type of composite is known as *polysilsesquioxane* or *oligosilsesquioxane* and has a general formula of $(RSiO_{1.5})_n$, where R is a hydrogen or carbon group. The base molecule is an inorganic nanomaterial cage made up of silicon atoms at the vertices and oxygen atoms at the middle of each edge. The organic groups are attached to the inorganic cage by covalent bonds to silicon atoms. In this way, the aggregation of polymer chains that are grafted to the cage is completely avoided or greatly reduced (see Figure 2.19).

The advantage of these composites is, on the one hand, the thermal and chemical stability of the inorganic material, which ensures the strength of the material. On the other hand, it is possible to modulate the functionality of organic groups to allow interesting physical properties (solubility, optical or electrical properties) to be obtained.

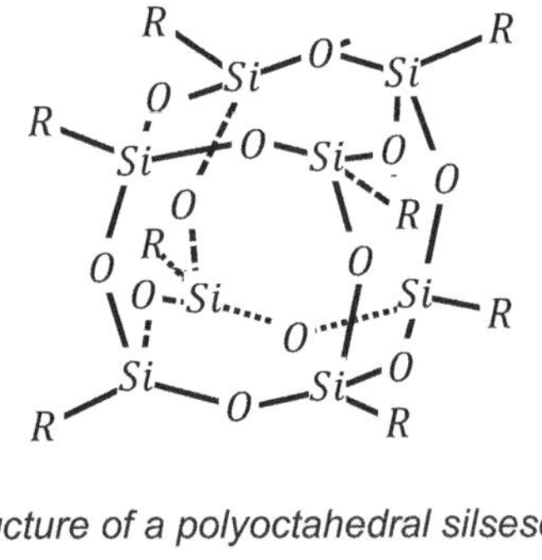

Figure 2.19. *Structure of a polyoctahedral silsesquioxane (POSS)*

2.7.3.4. In situ *polymerization method*

In this method, the polymerization is carried out directly on the surface of the nanoparticles. The polymer chains are formed in the presence of the nanoparticles previously and homogenously dispersed within the monomer solution. The polymerization process can be carried out on solid masses, in a solution or in a dispersed medium, depending on the characteristics of the polymer used. For example, nanocomposites of PANI/ZnO or $PANI/SnO_2$ are prepared by emulsion polymerization of aniline monomers in which the oxide particles are incorporated. Nanocomposites are made of a porous matrix of PANI containing nanoparticles of ZnO or SnO_2 (Alvi *et al.* 2010).

More recently, reduced graphene oxide (rGO) obtained through the oxidation of graphite is used as a matrix onto which aniline monomers are incorporated, and chains of PANI are grown by *in situ* polymerization (Vinoth *et al.* 2017). This technique is applied to obtain composites using carbon nanotubes with conjugated polymers, which are employed as materials for electrodes in devices to improve conductivity and adhesion to the electrode/active layer interface. For example, composites PEDOT-MWCNT-Pt are formed by polymerization of monomers EDOT in a solution containing multi-wall carbon nanotubes and nanoparticles of platinum adsorbed on the walls (Wang *et al.* 2011). These materials, when used in solar cells, allow their performance to be significantly improved.

2.7.3.5. *Use of nanostructured substrates*

The main difficulty in the preparation of composites is the dispersion in the nanoparticles or nanofillers in the matrix of the polymer material. Regardless of the technique used, a homogeneous dispersion of the particles is essential for ensuring that the material itself is homogeneous and has reliable physical characteristics. In practice, it is difficult to avoid aggregates when preparing solutions, as well as later when the composite material is deposited and shaped. To prevent this disadvantage, the polymer matrix is replaced with a matrix of inorganic materials, and the composite material is formed by filling the matrix with a polymer solution. The advantages of this technique are, first, the possibility of obtaining a homogeneous distribution of the nano-sized inorganic material, and second, the ability to precisely control the geometric size (length and diameter) of the composite material to be formed.

Nanostructures ZnO have been intensively studied for their use in solar cells. From a practical point of view, the nanowires or nanorods of ZnO are made on a substrate constituting a carpet on which the polymer films are deposited to constitute the nanocomposite layer (see Figure 2.20).

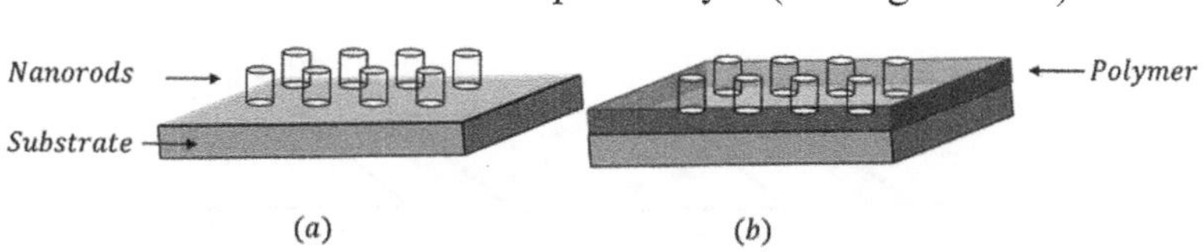

Figure 2.20. *Nanocomposites using a nanostructured substrate: (a) nanostructured substrate; b) nanocomposite. For a color version of this figure, see www.iste.co.uk/nguyen/electronics1.zip*

Note that the term "nanowires" refers to wires of nanometric size without any specific organization while the term "nanorods" refers to wires with larger diameters that are grown on a substrate without precise organization. The term "nanorod arrays" is used to refer to nanorods that are well organized (see Figure 2.21).

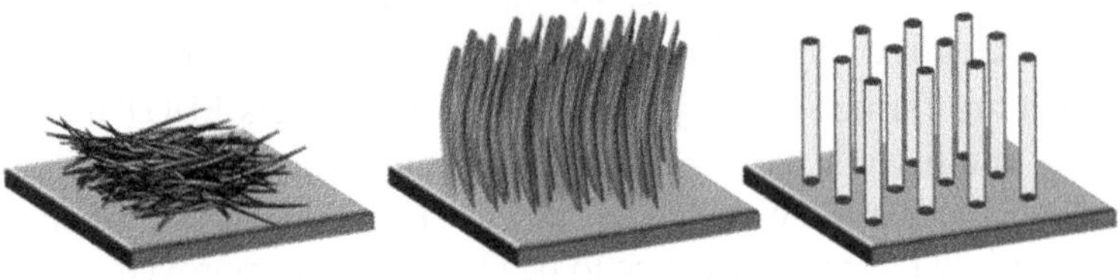

a) *b)* *c)*

Figure 2.21. *a) Nanowires; b) nanorods; c) nanorod arrays*

Several techniques can be used to fabricate nanorods (Yi *et al.* 2005):

– thermal evaporation of ZnO powder;

– *metal-organic chemical vapor deposition (MOCVD)* in the vapor phase: a thin film of ZnO is deposited on an alumina substrate under argon gas flow. Oxygen and dimethylzinc (DMZ) are introduced, and their reaction at high temperatures forms nanorods, which grow vertically at the surface of the substrate;

– hydrothermal synthesis: a buffer layer containing zinc acetate dispersed in a polymer matrix of PVA (poly(vinyl alcohol)) is deposited in a thin film on a substrate (see Figure 2.22). After being treated at high temperatures (~500°C), the polymer matrix is removed, and the nanocrystallites of ZnO are formed on the surface of the substrate. When annealed at ~95°C, the nanorods grow from the crystallites and form a lattice whose geometric size can be controlled.

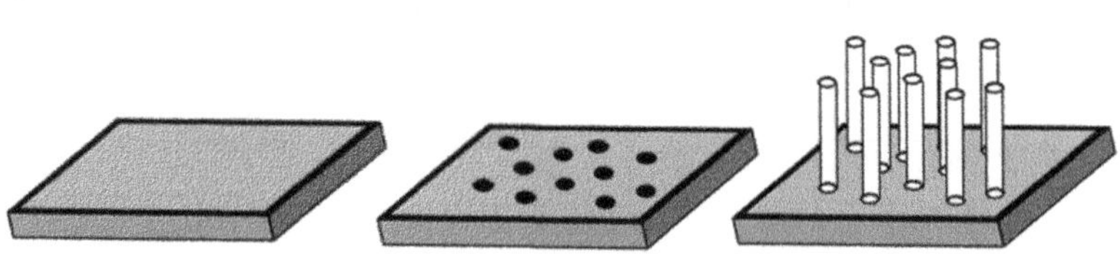

a) b) c)

Figure 2.22. *Hydrothermal fabrication of ZnO nanorods: a) buffer layer-covered substrate; b) formation of ZnO nanocrystallites; c) growth of nanorods from the nanocrystallites*

To properly control the nanostructured lattice, and particularly to ensure the orderly alignment of the nanorod arrays, it is necessary to use templates to place the bases of the nanowires before they grow.

2.8. Transparent and conductive materials

The technology of the devices intended to be used as displays requires the use of both conductive and transparent materials in order to be able to connect them to external electrical sources (power supplies, batteries) and to allow light to pass into or out from the component towards the external medium. These particular materials are often oxides or SCs, referred to as *transparent conducting oxides* (TCOs). The most commonly used material for the devices is *indium-doped tin oxide* (ITO), but there are also other materials that have been studied for replacing ITO in the future. The main characteristics to be expected from these materials are good transparency with a transmittance close to 90%, good electrical conductivity of about 10^4 S/cm and a low surface roughness of less than 5 nm. It is worth noting that, from an electrical point of view, the resistivity of *thin films* is usually expressed in terms of *sheet resistance* or *sheet resistivity*, defined by the relation:

$$\rho_s = \frac{\rho}{d} \qquad [2.8]$$

where ρ is the resistivity and d is the thickness of the film. The relation ρ/d is expressed in unit of resistance and is denoted as Ω/□ (ohms per square). For TCOs, the sheet resistance must be less than 100 Ω/□.

2.8.1. *Indium tin oxide*

ITO is currently widely used in industry for manufacturing optoelectronic devices available on the market. Before it was developed, Au and Ag semi-transparent metal films were used, the transmittance of which could reach ~50% for transparent and electrically conductive layers. Thin films of ITO have remarkable optical and electrical properties: they have good electrical conductivity ($\sim 10^4\ \mathrm{S\ cm^{-1}}$) like some metals and are as transparent as some insulators (with a visible light transmittance close to 80%). On the other hand, despite the high electrical conductivity of ITO, its thermal emissivity is relatively low, unlike metals, and therefore can serve as a filter for insulating windows.

ITO thin films can be deposited by several techniques, and the properties of the films obtained vary depending on the chemical composition and method of preparation. The films are usually deposited on glass slides, and the methods often used are sputter deposition, chemical vapor deposition, pulsed laser deposition and spray pyrolysis. Note that there is no standard for ITO, and commercial products often have characteristics that are quite different. It has been shown that the quality of the ITO substrate influences the performance of devices such as OLEDs or OSCs. For the use of these substrates in these devices, it is necessary to treat the surface of the ITO layer before the deposition of the organic films. The treatments used are intended to stabilize the structure of the oxide and smooth the surface of the film. Indeed, it can firstly be observed that ITO can decompose when in contact with the organic film during its deposition, and the metal ions of the oxide can diffuse into the components, causing premature degradation. Second, the analysis of the morphology of the film after deposition shows a rough appearance with material peaks of a high intensity (~100 nm). These peaks can be responsible for short circuits in components in which the average thickness of the active layer is about 100 nm. Other modifications of the surface of the film, including the surface bonds and the formation of ions, promote the injection of charges from the electrode into the active layers. The effects that are generated depend on the treatment used. These treatments include heat treatment (Nunes De Carvalho *et al.* 2000), optical treatment (UV) (Hu *et al.* 2009), plasma treatment (Milliron *et al.* 2000) and chemical treatment (Nguyen *et al.* 2003).

The fact that indium is a scarce resource means that ITO is costly to produce, in addition to its variable quality problems and the instability of the oxide during its use. Despite these disadvantages, it is almost systematically used in optoelectronic components, because the current substitute materials do not possess optical and electrical qualities comparable to those of ITO.

2.8.2. *Fluorine-doped tin oxide*

Fluorine-doped tin oxide is obtained by doping tin dioxide SnO_2 with fluorine atoms, which replace the oxygen sites in the lattice. The fluorine atom is an electron donor, and due to its size comparable to that of the oxygen atom, fluorine doping does not affect the structural stability of the tin oxide. The electrical conductivity (~ 10^3 S cm^{-1}) is less than that of ITO, and the visible light transmittance is close to 80%. The optical gap of the FTO is ~4 eV, and its use as an electrode in solar cells improves the absorption of UV radiation. The advantages of this TCO are its simple deposition process, low production cost and the thermal stability of the material, these qualities being what allows it to be used in manufacturing organic solar cells.

2.8.3. *Other transparent oxide conductors*

Several TCOs have been developed and studied for replacing ITO in optoelectronic applications in the near future. Good electrical and optical performance has been achieved in these materials, and the criteria for their use as an electrode in optoelectronic devices are satisfactory. Nevertheless, synthesizing these materials still requires further improvements in order to achieve the level of reliability of ITO, which would explain why very few TCO materials are used commercially in producing the electrode films in devices. Among the most promising materials other than FTO, doped zinc oxides (ZnO) possess interesting properties, particularly the fact they are economical to produce, since zinc is abundant and non-toxic. Aluminum-doped zinc oxide (AZO) has good electrical conductivity (> 10^4 S cm^{-1}) and good transmittance (~90%) due to the wide band gap of ZnO. In addition, compared to ITO, this oxide is more thermally and chemically stable at high temperatures. The material's weak point is its aging and the fact that it is difficult to control its chemical composition.

2.8.4. *Other transparent conductive materials*

In organic electronics, it is preferable to use flexible materials in order to take advantage of and enhance the flexibility of devices. It is clear that rigid elements, such as oxides, are less suitable for use as electrodes than polymers are for these components. However, there are few organic materials capable of being satisfactorily doped while at the same time remaining stable for use in devices. Other conductive nanomaterials can also be formed in lattices and used as transparent electrodes with good optical and electrical quality.

2.8.4.1. *PEDOT:PSS*

PEDOT:PSS is a mixture of two polymers: poly(3.4-ethylene dioxythiophene) (PEDOT) and sodium poly(styrene sulfonate) (PSS). This polymer is commonly used in many organic electronic components as electrodes, transport layers or dopants. PEDOT is obtained through the polymerization of EDOT monomers, and in its oxidized state, it contains protonated sulfur atoms with a positive charge. When mixed with PSS, negatively charged sulfonated polystyrene containing sulfonate SO_3 groups with an Na^+ sodium ion, the introduced charges form counter-ions for dispersing the chains of PEDOT in the oxidized (doped) state in an aqueous medium. In this way, a polymer solution is obtained that can be deposited in thin films with the usual deposition methods. PEDOT:PSS films have a good electrical conductivity ($\sim 10^3$ S cm^{-1}) and a good transparency that varies depending on the thickness of the film (~90% for a thickness of 100 nm), which makes it possible to use as a conductive transparent electrode. These films are frequently deposited on the ITO layer and are used as a hole transport layer in the devices. These devices show a significant improvement in performance compared with those that do not contain PEDOT:PSS. However, it can be noted that the interface between ITO/PEDOT:PSS contains defects or traps that alter the stability of the structure (diffusion of metal atoms of the ITO in the active layer (De Jong *et al.* 2000)) and the carrier transport (trapping) (Nguyen and Renaud 2009). Treatments of the polymer will be necessary to improve the stability and conductivity of PEDOT:PSS for its use in devices (Nguyen *et al.* 2003).

2.8.4.2. *Networks of conductive nanomaterials*

These materials consist of conductive nanomaterials assembled as an organized or non-organized surface and thin enough to be optically transparent. The nanomaterials can be metal nanowires or carbon nanomaterials (carbon nanotubes or graphene).

2.8.4.2.1. Metal nanowires

To obtain transparent metal films, a metal grid can be made with the conductive lines designed to allow sufficient light to pass through. Several techniques (lithography, inkjet printing, evaporation) have made it possible to create transparent metal grids to be used as electrodes, but production costs are high. Metal nanowires (made from silver or copper), with an average diameter between 50 and 100 nm (see Figure 2.23), make transparent metal electrodes cheaper and use less material. To obtain a conductive film, the nanowire network must exceed the *percolation threshold* with a minimum of materials in order to preserve the transparency of the film.

Several deposition techniques can be used to make these films, including spin-coating, spray-coating and printing. In any case, the nanowires are incorporated into an aqueous solution to facilitate the deposition, and the deposited film is treated to retain only the lattice of metal materials.

The morphology of the films remains a problem because the high roughness values can cause short circuits. This lattice can be used as is or incorporated into a PEDOT:PSS film to obtain a composite, which will form a conductive and transparent electrode with an acceptable morphology.

a) b)

Figure 2.23. *Transparent and conductive electrodes: a) unordered array of metal nanowires; b) ordered array of metal nanowires*

2.8.4.2.2. Carbon-based nanomaterials

Instead of metal nanowires, nanomaterials such as carbon nanotubes or graphene are deposited in thin layers. Since these materials are excellent conductors, the obtained layers have good electrical conductivity. The preparation technique allows transparent films to be obtained (with transmittance of ~90%).

To produce an electrode with CNTs, the tubes are incorporated into a surfactant solution, which is thoroughly mixed to obtain a homogeneous dispersion of the tubes. This solution is then filtered through a membrane, and a thin layer of CNTs is formed on the surface of the membrane. This layer is transferred to the surface of a glass or plastic substrate using a poly(dimethysiloxane) PDMS stamp (Zhang *et al.* 2006). The sheet resistance of the best layer is 160 Ω/□ with a transmittance of 87%.

The graphene used as a transparent electrode is deposited from an aqueous solution of oxidized graphene. The film obtained is then reduced by heat treatment, and the optical and electrical performance is low since the reduction of the oxide alters the structure of graphene. Higher quality in the film can be achieved by depositing graphene on a metal substrate (Cu) using the roll-to-roll chemical vapor deposition (CVD) method (Burrows *et al.* 1994). The graphene/Cu film is then transferred to a substrate of polyethylene terephthalate (PET), and the metal layer is dissolved. The contact resistance is rather high (~500 Ω/□), and the transmittance is more than 80%. The weakness of graphene films is their high defect concentration, which is difficult to control.

2.9. Materials for encapsulation

When organic electronic devices operate in air without protection, they are degraded. Degradation occurs in the organic layer and also at the electrodes, which reduces the lifetime of the components. To protect the device and extend its lifetime, an encapsulation layer must be used. The material used for encapsulation must be as impervious as possible to the diffusion of gases (oxygen or water vapor) with kinetics characterized by a parameter called the *transmission rate* (TR) expressed in $g.m^{-2}.day^{-1}$. We will use the parameter OTR (oxygen transmission rate) for the diffusion of oxygen, and WVTR (water vapor transmission rate) for water vapor. The gas diffusion in materials is made easier by the presence of defects that create propagation pathways.

For effective protection of organic devices, the material must have a transmission rate of lower than $10^{-4} g.m^{-2}.day^{-1}$ for solar cells and $10^{-6} g.m^{-2}.day^{-1}$ for light-emitting diodes. On the other hand, as the encapsulation film is deposited on the device, it must be optically transparent, and as flexible and light as possible to preserve the qualities of the component. These materials will be referred to as *high barrier* (HB) or *ultra-high barrier* (UHB) materials.

2.9.1. *Glass slides*

Glass slides provide for good encapsulation for devices manufactured in laboratories because their use is simple and relatively efficient (Weaver *et al.* 2002). The blade is set in place through pasting along the contour of the substrate using an adhesive (such as epoxy glue, benzocyclobutene (BCB) or parylene) in inert atmosphere or a vacuum (see Figure 2.24a). The protected diodes will operate continuously for 10,000 hours, despite a loss of 60% of the initial luminance. The use of a glass slide leads to less flexibility of the devices and cannot be adapted to the manufacturing and use of components for commercial sale.

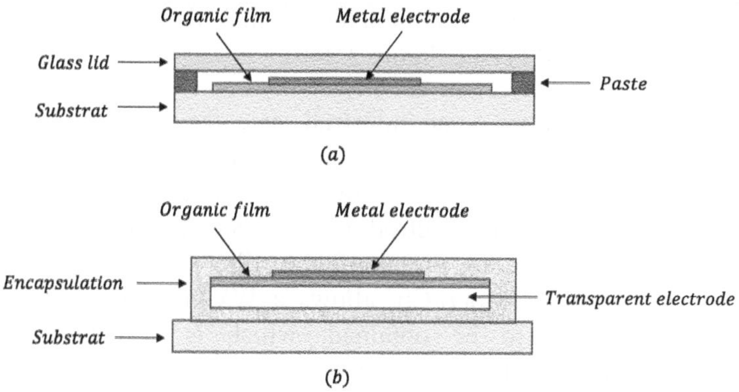

Figure 2.24. *Encapsulation of organic electronic devices: a) by a glass slide; b) by thin multilayered films. For a color version of this figure, see www.iste.co.uk/nguyen/electronics1.zip*

2.9.2. *Hybrid multilayers*

To preserve the flexibility of the component, an organic film is used as an encapsulation layer. However, there is no organic material effective enough to stop the diffusion of oxygen or water vapor, which degrades the components. Therefore, to achieve encapsulations of HB or UHB, hybrid materials composed of organic layers alternating with inorganic layers will be used. This alternation allows the path of the diffusion flow to be lengthened and slows the propagation of gases in the device, and thus extends the stability of the electronic device. A pair of alternating layers is called a *dyad*.

Hybrid layers of aluminum oxide Al_2O_3/polyacrylate have been used to efficiently encapsulate light-emitting diodes with a water vapor transmission rate less than $2 \times 10^{-6} g.m^{-2}.day^{-1}$ (Weaver *et al.* 2002). The number of dyads used to form a barrier film cover varies between three and five.

This technique is effective, yet it still has drawbacks to its implementation in practice. In reality, the deposition of alternating layers is a long and expensive process. Furthermore, the oxide layers are fragile and can be damaged by a repeated bending of the substrate.

It should also be noted that the diffusion of gases is carried out not only through the encapsulation layer but also through the substrate, and through the edges of the layers of the device (lateral permeation). It is therefore necessary for the device to be completely coated, then deposited onto the substrate (Park *et al.* 2011) (see Figure 2.24b).

To create encapsulations on an industrial scale, the *roll-to-roll lamination* technique is used. This consists of gluing a barrier film directly to the substrate with an adhesive, onto which the electronic component (transparent electrode/active layer/metal electrode) has been printed on the same production line. The use of an aluminum film of thickness 100 nm, reinforced by a layer of MoO_3/Cu, allows a WVTR transmission rate of $1 \times 10^{-6} g.m^{-2}.day^{-1}$ to be obtained, which is acceptable for low production costs (Dollinger *et al.* 2017).

3

Optical Processes

3.1. Introduction

The main applications of organic electronic devices are *organic light-emitting diodes* (OLEDs) and *solar cells*, or organic photovoltaics (OPVs). The acronym OSC (organic solar cell) is also used. These devices make use of optical processes occurring in organic materials that interact with light radiation. In this chapter, we will present the main processes that will enable the operation of these devices to be better understood.

3.2. Interaction between light and molecules

3.2.1. *Electronic transitions*

When an organic material is exposed to light, the molecules of the material are subjected to the electric field of the photons. When affected by this field, electrons can make *transitions* between different energy levels. The energy ΔE of an electronic transition is the energy supplied by a radiation of frequency v:

$$\Delta E = hv \qquad [3.1]$$

where h is Planck's constant.

The electronic transition changes the state of the molecule from a fundamental state (or lowest energy state) to an excited state (or higher energy state). Conversely, in the same material, the electronic transition from the excited state to the fundamental state can also occur by yielding an energy in a form that depends on the conditions under which the transition occurs.

A material has several characteristic energy states, which reflect its molecular structure and its own unique energy processes. Figure 3.1 shows the energy levels of a diatomic molecule as a function of the intermolecular distance (called the Morse function). The fundamental (or ground) state represents the equilibrium of the two atoms separated by a distance of R_e. When the molecule is excited, the atoms begin to move far away from each other. As the distance between two atoms increases, the inter-atomic bonds gradually weaken, and the potential binding energy of the atoms approaches a limit. As a result, the difference between energy levels decreases (see Figure 3.1a).

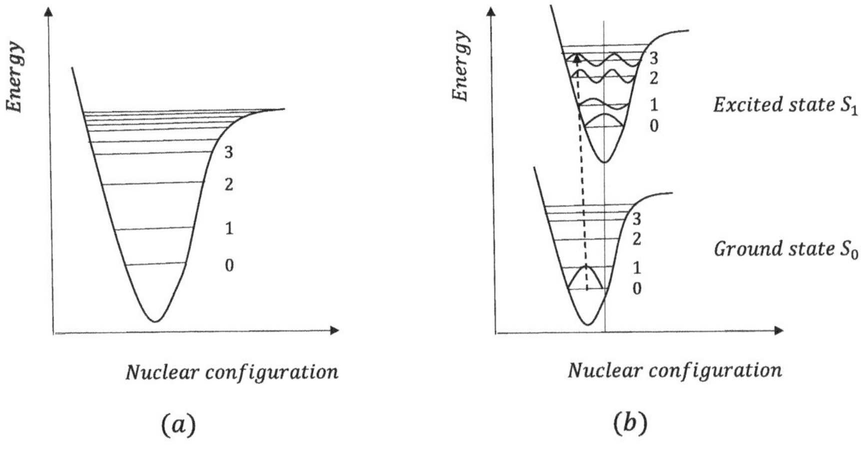

Figure 3.1. *Potential energy of a diatomic molecule as a function of the inter-atomic distance: a) Morse potential of the isolated molecule; b) Morse potential of the molecule at the fundamental state S_0 and the excited singlet state S_1*

3.2.2. *Selection rules*

There are two selection rules for electronic transitions in a material. The first rule is called the spin rule that relates to spin-forbidden transitions. The second rule is called the orbital rule that relates to symmetry-forbidden transitions, which will not be treated in this section.

3.2.2.1. *Spin-forbidden transitions*

This rule states that, during an electronic transition, the electron conserves the same orientation of its spin. As a result, transitions between electronic states of different multiplicities of spin are forbidden. In other words, a transition between a singlet state of multiplicity $M = 1$ and a triplet state of multiplicity $M = 3$ is not allowed, while a transition between a singlet state and the fundamental state of multiplicity $M = 1$ is allowed. It should be recalled that the multiplicity is defined by the relationship:

$$M = 2 \times |s_1 + s_2| + 1 = 2 \times S + 1 \qquad [3.2]$$

where s_1 and s_2 are the two values of parallel and antiparallel spins and S is the total spin. For two antiparallel spins, the total spin is $S = \left|\frac{1}{2} - \frac{1}{2}\right| = 0$ and for two parallel spins $S = \left|\frac{1}{2} + \frac{1}{2}\right| = 1$.

Although the rule theoretically prohibits transitions between singlet and triplet states, such transitions are observed in practice, because the wave functions can partially overlap through spin–orbit coupling, making the transition between these states possible. These exchanges are generally limited and the corresponding process is low in intensity.

3.2.2.2. *The Franck–Condon principle*

This principle states that, during an electronic transition, the positions of the nuclei of the molecules do not change with respect to their positions before the transition, since the electronic transitions are much faster than nuclei transitions. In other words, the electronic state of the excited molecule has the same geometric configuration as that of its fundamental state, and the probability of transition between two states will be even more likely as the vibrational wave functions of the corresponding states overlap significantly.

As a result, the transition between two energy states is vertical, which occurs between certain energy states having favorable configurations, that is, with a minimum change in nuclear configuration (see Figure 3.1b). The most likely transition in this example will take place between the levels $v = 0$ of state S_1 and $v = 1$ of state S_0, which is indicated by the vertical arrow.

The rules of selection apply to electronic transitions between the fundamental state and the excited states of the molecule in both directions (from the fundamental state to the excited state or from the excited state to the fundamental state) and have important consequences on optical processes, as we will see below.

3.3. Optical processes

3.3.1. *Light absorption*

3.3.1.1. *Absorption spectrum*

The effects of the electromagnetic field of light cause the charges of the illuminated material to move and oscillate at a characteristic frequency ν_i, which depends on the nature of the material. When this frequency is close to the frequency of light, it is absorbed by the material, and the intensity of radiation passing through it decreases.

The process of the absorption of radiation by the material follows the Beer–Lambert law, which considers three factors affecting the light intensity: 1) the amount of material that absorbs the light, 2) the path or distance traveled by the light and 3) the probability that a photon of a given energy is absorbed by the material.

Consider a light beam of an intensity I_0 that comes into contact with the surface of a material in the *Ox* direction perpendicular to the surface. Due to absorption, its intensity $I(x)$ decreases as it enters the medium. The change in the intensity of the radiation as a function of the distance of penetration x is written following the Beer–Lambert law:

$$\frac{dI}{I} = -\alpha c dx \qquad [3.3]$$

where α is a constant and c is the concentration of absorbing molecules.

By integrating this relationship over the path of length l, we obtain:

$$\ln\frac{I_0}{I} = \alpha c l \qquad [3.4]$$

where I_0 and I are the radiation intensities at distances $x = 0$ and $x = l$.

We can also write:

$$I = I_0 exp(-\alpha cl) \quad [3.5]$$

Absorbance or optical density is defined as the ability of the material to absorb light at a wavelength λ following the expression:

$$A(\lambda) = \log\left(\frac{I_0(\lambda)}{I(\lambda)}\right) = \varepsilon(\lambda)cl \quad [3.6]$$

where $\varepsilon(\lambda) = \frac{\alpha}{ln(10)} = \frac{\alpha}{2,303}$ is the molar absorption coefficient, expressed in $Lmol^{-1}cm^{-1}$ or $m^3mol^{-1}cm^{-1}$, c is the molar concentration and l is the thickness of the material.

The absorption spectrum of the material represents the change of $A(\lambda)$ or $\varepsilon(\lambda)$ as a function of the wavelength λ of the incident light. It contains absorption bands that correspond to the electronic transitions that have occurred between states in the material. Each band has a maximum corresponding to the wavelength λ_{max}. For low-energy electronic transitions, the absorption band is observed in the range of long wavelengths, and vice versa: the absorption band of high energy transitions is located in the range of short wavelengths.

The absorption of a material composed of several elements is the sum of the contributions of each of the elements. Thus, the use of composites made up of materials, which absorb light of wavelengths outside the visible spectrum, allows us to obtain a wide absorption spectrum. This is advantageous for the operation of devices such as solar cells.

3.3.1.2. *Electronic transitions*

The absorption of photons by the material creates electronic transitions from the fundamental level S_0 to the higher energy levels, denoted as $S_1, S_2, \ldots, S_n$, depending on the transition probability and the energy of photons (i.e. the wavelength λ of light). The greater this energy, the higher the level that is reached. These transitions will also be denoted as $0 \rightarrow n$ to denote the level for the outgoing energy (S_0) and the incoming energy (S_n). The excited molecules quickly return from this energy level to the first singlet level S_1, losing energy through collisions (thermalization). The electronic transition is $S_n \rightarrow S_1$ and the process is called *internal conversion*.

Transitions between triplet states are also possible when an excited molecule in the state T_1 absorbs a photon of a different wavelength, allowing it to reach a higher energy level. The electronic transition $T_1 \rightarrow T_n$ is allowed under the selection spin rule.

3.3.1.3. *Effects of aggregates*

Aggregates are assemblies of molecules that are formed by electrostatic interactions in materials in a condensed state. They can be found in concentrated solutions or in thin organic films. When an aggregate is optically excited, the absorption of photons is not systematically equal to the sum of the responses from each of the molecules forming the aggregate, but depends on how the molecules are organized. This observation is based on the shift of the maximum of the absorption spectrum of the polymer that contains aggregates with respect to the maximum of the monomer spectrum.

For example, in the case of a dimer formed from two monomers, the molecules can have a stacking (or sandwich-type) arrangement or an alignment arrangement. Each organization type corresponds to an absorption band that is shifted from the absorption band of the monomer with the same chemical nature. The absorption band of energy that is higher than that of the monomer band corresponds to the stacking arrangement of the molecules. These aggregates are of type H. Conversely, the aggregates whose molecules are arranged in alignment are of type J and their absorption band energy is lower than that of the corresponding monomer. The difference in absorption between two types of aggregates originates from the energy level of the excited state. The parallel plane alignment or stacking of molecules corresponds to the highest moment of transition, while the head-to-tail alignment offers the lowest energy (see Figure 3.2). Aggregates of types H and J are formed according to the inclination α of their dipole moment relative to the longest axis of the aggregate. If $\alpha > 54,7°$, H aggregates are formed, and if $\alpha > 54,7°$, J aggregates are formed. A study of the absorption spectrum allows us to determine the type of aggregates formed in polymer films. For example, in poly(hexylthiophene), several aggregate configurations can be formed depending on the synthesis method (Spano *et al.* 2009).

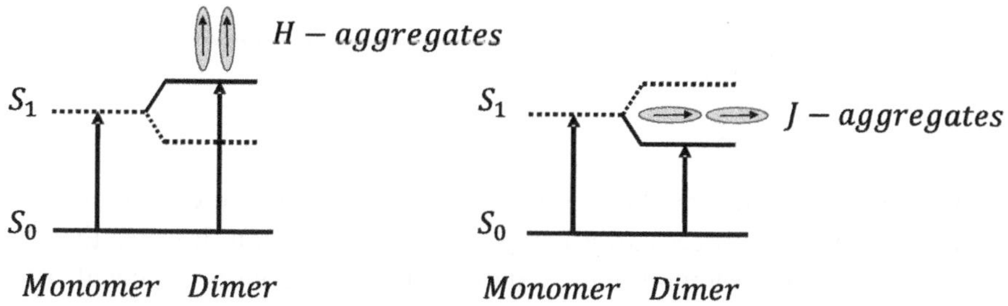

Figure 3.2. *Diagram of the energy and electronic transitions in the process of absorption by aggregates of types H and J*

When a polymer film contains aggregates, the original absorption band of the material will split into two or more bands, with each of them relating to the electronic transitions towards the levels created by the interactions between molecules. The use of different solvents allows us to promote the formation of a given type of aggregate and to study this formation by recording the absorption spectra. In fact, in absorption spectroscopy, there is a particular wavelength at which all species or chemical components have the same molar absorbance, that is, the same *extinction coefficient*. This wavelength is called the *isosbestic wavelength* or the *isosbestic point*. At this point, the absorption spectra of two species cross each other and have the same absorbance. The presence of an isosbestic point in the absorption spectra suggests that there is a balance between the species and that it can be exploited to study the formation of aggregation states in polymer films that affect their photophysical properties.

When aggregates of type H or J are formed in polymers, the energy changes in the electronic transitions lead to a shift in the absorption spectra of the material without aggregates. Thus, we can observe either a blue shift of the spectrum (for H-type aggregates) or a red shift (for J-type aggregates). However, other factors can affect the transition energy in the material, and relying solely on the shift of the spectrum would not allow us to determine the presence of aggregates with certainty and distinguish them. One method of identifying aggregates was proposed by Spano (Spano and Silva 2014), which consists of studying the intensity ratio of two lowest energy transitions (A_{0-0} and A_{0-1}) when the aggregation of molecules takes place. For H-type aggregates, the ratio of intensities A_{0-0} / A_{0-1} decreases as the aggregates increase, due to the exciton coupling between the molecules.

3.3.2. *Light emission*

3.3.2.1. *Kasha's rule*

This empirical rule applies to all materials, with a few exceptions, for their light emission process after being excited by radiation. It states that the radiative emission following the excitation of the material is expected from the electronic transitions from the lowest excited state to the fundamental state. In other words, only the transitions $S_1 \rightarrow S_0$ of the singlet states allow for the material to yield radiative emission.

3.3.2.2. *Emission from a singlet state*

When the optical excitation has ceased and after a possible internal conversion, the molecule will return from the singlet state S_1 to the fundamental state S_0 through the electronic transition $S_1 \rightarrow S_0$. This direct transition is carried out over a very short time, of the order of 10^{-8} s. If the transition $S_1 \rightarrow S_0$ is *radiative*, the energy in play is transformed into the emission of a photon of energy equal to the energy difference between these two levels. In SCs, this energy is equal to the gap value, such that for an emission of photons of a frequency ν, we obtain:

$$E_G = h\nu \qquad [3.7]$$

The transition $S_1 \rightarrow S_0$ can also be *non-radiative*, and in this case, the energy in play is transformed into thermal energy that is dissipated in the molecule.

The process of electronic transition $S_1 \rightarrow S_0$ involving the emission of a photon is called *fluorescence* and the process that does not involve any emission is called non-radiative de-excitation or *quenching*.

3.3.2.3. *Emission from a triplet state*

The transition $S_1 \rightarrow S_0$ can also be carried out indirectly through triplet states. In this mechanism, the first electronic transition takes place between the states of S_1 and T_1. This is possible because the first energy level of the triplets T_1 is generally lower than that of the excited singlet state S_1. The transition between states S_1 and T_1 is called *intersystem crossing*. It is non-radiative and allows the molecule to pass from the excited vibrational state S_1 to the excited vibrational state T_1, states which have two different spins. From T_1, the molecule returns to the fundamental state S_0 either with the emission of a photon (a process called *phosphorescence*) or without any emission (called the quenching process). Phosphorescence is an energy state transition that is normally forbidden due to its occurrence between two different states of spin, but can become possible because of spin–orbit coupling. It requires a change in the state of spin, and thus the transitions $T_1 \rightarrow S_0$ are long and occur rarely. The duration of the emission can reach $10^{-2} - 10^{3}$ s.

3.3.2.4. *Delayed fluorescence process*

The normal emission of photons by transition $S_1 \rightarrow S_0$ is a spontaneous emission, known as *fluorescence*. Another process exists, known as *delayed fluorescence*, which is similar to spontaneous fluorescence but has a lifetime that is longer and comparable to that of phosphorescence. This process originates from the repopulation of the singlet states S_1 by the triplet states T_1 through an inverted *intersystem crossing* process by the transition $T_1 \rightarrow S_1$. In this case, two triplet states are fused to form a state with an energy level higher than that of the singlet state S_1, thus allowing the transfer between the two levels ($2E_{T1} > E_{S1}$). The fusion of the two triplet states is called *triplet–triplet annihilation* (TTA). This process is called *P-type delayed fluorescence*. Another transition process $T_1 \rightarrow S_1$ can take place in the material when the energy difference between the levels T_1 and S_1 is small and the lifetime of T_1 is sufficiently long. The delayed fluorescence that occurs due to this process is of type E.

3.3.2.5. *Mirror effect and Stokes shift*

The 0–0 transition involves the vibrational levels of 0 of each of the states S_0 and S_1, which is the same for the processes of absorption and fluorescence of a material. In the spectra of absorption and fluorescence, there is a line called the *zero-phonon line* that corresponds to this transition, and which is the most intense of the spectrum. The other lines are symmetrical with respect to the zero-phonon line in wavelength, and the shape of the spectra is also symmetrical. This characteristic is known as the *mirror symmetry* of absorption and fluorescence spectra. In practice, it is observed that the fluorescence spectrum is located at higher wavelengths (with lower energy) than the absorption spectrum due to an energy loss by vibrational relaxation from the excited state. The *Stokes rule* states that the wavelength of fluorescence emission of a given material is always higher than that of the absorption of that same material. The process of energy loss can be explained as follows: in the excited state, the electron has an energy that is located at a given vibronic energy level. During the de-excitation process, it passes to a lower vibronic energy level, and energy is dissipated as heat (thermalization). When the electron returns to the fundamental state by a radiative transition, its final energy is lower than its initial value, and therefore the emitted photon will have a higher wavelength than the one that creates the absorption. The difference between the maxima of the frequencies or wavelengths of the absorption and the emission spectra is called the *Stokes shift*:

$$\Delta\nu_{Stokes} = \nu_{max}^{A} - \nu_{max}^{E} \qquad [3.8]$$

Generally, a partial overlap of the emission and absorption spectra is observed in the materials, suggesting that photons can be emitted at wavelengths shorter than those of the absorption spectrum, contradicting the Stokes rule. These are actually the transitions occurring due to the residual energy states located above level 0 of the ground and excited states, which are created by thermal agitation, and the phenomenon of overlap will not be observed if the spectra are recorded at low temperatures.

3.3.3. *Perrin–Jablonski diagram*

The electronic transitions between the energy levels of a molecule and the optical processes involved in these transitions are schematically summarized in Figure 3.3. This diagram is called the *Perrin–Jablonski diagram*.

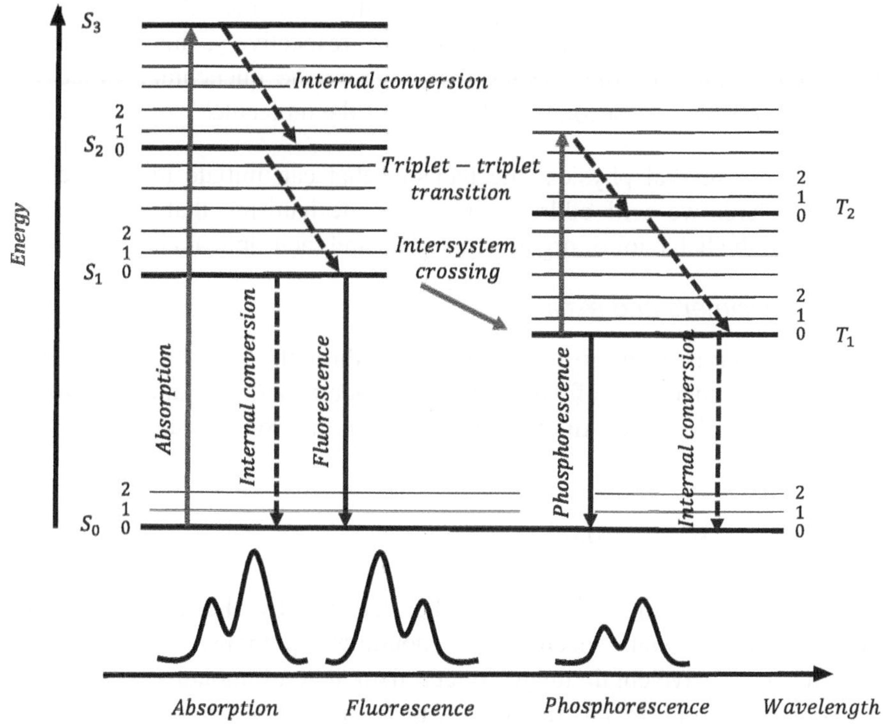

Figure 3.3. *Energy diagram and possible electronic transitions in a molecule according to the Perrin–Jablonski model. For a color version of this figure, see www.iste.co.uk/nguyen/electronics1.zip*

3.3.4. *Quenching*

An excited electronic state can be relaxed to its fundamental level without the emission of photons. This process is known as *quenching* ($S_1 \rightarrow S_0$ or $T_1 \rightarrow S_0$ internal conversion) in the Perrin–Jablonski diagram.

The quenching reaction of the process can be described using the following equation:

$$M^* + Q \rightarrow M\left(\text{or}\, M^{**}\right) + Q\left(\text{or}\, Q^*\right) \quad [3.9]$$

where M and M^* represent, respectively, the fundamental state and the excited state of the molecule; Q and Q^* are, respectively, the fundamental and excited states of the species that triggers the quenching, called the *quencher*; and M^{**} is a second excited state of the molecule.

There are several physical mechanisms that can initiate the quenching process. Here, we will address the main mechanisms that allow us to understand the behavior of organic materials and their optical characteristics.

3.3.4.1. *Excimers and exciplexes*

An *excimer* is a dimer molecule formed through the interaction of two identical monomers, one of which is in the fundamental state and the other in the excited singlet state. An excimer is itself in the excited state, as given in the equation:

$${}^1M^* + {}^1M \rightarrow {}^1(MM)^* \quad [3.10]$$

The formation of excimers can be detected by the changes in the fluorescence emission spectrum of monomer solutions. Excimers emit photons in a wavelength range longer than that of monomers and their emission spectrum is unstructured. Since the formation of excimers is dependent on the concentration of monomers, the intensity of the excimer emission spectrum increases as the concentration of monomers increases. At the same time, since their formation leads to the quenching of monomers, the emission spectrum of monomers decreases in intensity as their concentration increases.

An *exciplex* is a complex in the excited state formed by the interaction of an electron donor molecule in the excited (or fundamental) state with an electron acceptor molecule in the fundamental or (excited) state.

As with conventional SCs, some electron donor molecules are able to give electrons to electron acceptor molecules. These are denoted as D and A, respectively. The formation of an exciplex is described by the following equations:

$$^{1}D^{*} + \mathrm{A} \rightleftarrows {}^{1}(DA)^{*} \qquad [3.11a]$$

$$D + {}^{1}A^{*} \rightleftarrows {}^{1}(DA)^{*} \qquad [3.11b]$$

The effect of the formation of exciplexes on the fluorescence emission of monomers is similar to that of excimers, that is, an increase in the intensity of the emission band in the red region and a decrease in the intensity of the emission band of monomers. A unique characteristic of exciplexes, which distinguishes them from excimers, is the possibility of formation with triplet states.

3.3.4.2. *Resonance energy transfer*

The fluorescence quenching process can be initiated by energy transfer between molecules in the material. It takes place between an excited molecule D^{*} (donor) and a molecule initially at rest in the fundamental state A (acceptor).

In the classical theory of energy transfer, the process is described as the interaction between two oscillating electric dipoles. The electron of the donor molecule D produces an oscillating electric field that starts *resonating* with the electrons of the acceptor molecule A, which, in turn, also oscillate. The molecule A is excited and the process will repeat among the nearby molecules, resulting in an energy transfer in the material.

The energy transfer occurs due to the dipole–dipole interaction with dipole moments μ_1 and μ_2 of donor and acceptor molecules, respectively.

For two different molecules D and A of different natures, the transfer reaction, called a *heterotransfer*, is written as:

$$D^{*} + A \rightarrow D + A^{*} \qquad [3.12]$$

For two identical molecules, the transfer reaction, called a *homotransfer*, is written as:

$$D^{*} + D \rightarrow D + D^{*} \qquad [3.13]$$

The energy transfer process is possible in any material, provided that the emission spectrum of the donor molecule *D* and the absorption spectrum of the acceptor molecule *A* overlap (see Figure 3.4). In other words, a fraction of the photons emitted by *D* must be reabsorbed by *A* in the same wavelength domain.

Two types of energy transfer can be distinguished: *radiative transfer* and *non-radiative transfer*.

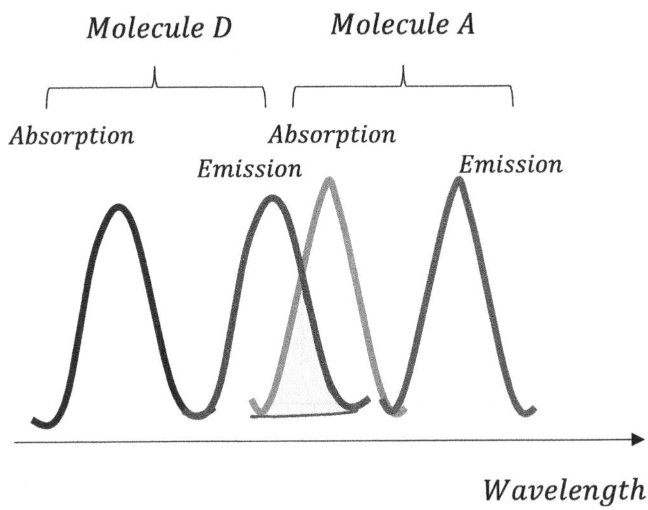

Figure 3.4. *Overlap of the emission spectrum of the donor molecule D and the absorption spectrum of the acceptor molecule A. For a color version of this figure, see www.iste.co.uk/nguyen/electronics1.zip*

3.3.4.2.1. Radiative energy transfer

In this process, the energy $h\nu$ of the photons emitted by the molecule *D* as it de-excites is absorbed by the molecule *A*, which becomes excited. After de-exciting, the molecule *A* returns to its fundamental state by emitting photons of an energy $h\nu'$ lower than the initial energy of the molecule *D*.

The transfer equations are as follows:

$$D^* \rightarrow D + h\nu \qquad [3.14a]$$

$$A + h\nu \rightarrow A^* \qquad [3.14b]$$

Since the energy of the molecule D decreases with the transfer, the intensity of its emission spectrum decreases in the overlap region of the absorption and emission spectra.

3.3.4.2.2. Non-radiative energy transfer

The energy transfer between molecules occurs through two types of interactions: the Coulombic interaction between long-distance dipoles (or the Förster mechanism), and the overlap of molecular orbitals at a short distance (or the Dexter mechanism). In both cases, the transfer takes place between a donor in the excited state and an acceptor in the fundamental state. The total energy involved is the sum of two terms: the Coulombic interaction energy U_C and the exchange energy U_E. The Coulomb potential energy is the transfer energy between the molecules when the electron in the excited state D^* returns to the fundamental state D and, simultaneously, the electron in the fundamental state A is promoted to the excited state A^* (see Figure 3.5a). The exchange energy is the transfer energy of the exchange process of two electrons between the states A and D and the states D^* and A^* (see Figure 3.5b).

In the Förster mechanism, the transfer is a resonant process that occurs between two moments μ_1 (of the process $D^* \rightarrow D$) and μ_2 (of the process $A \rightarrow A^*$) of the excited molecule D^* and the molecule A at rest according to equation [3.11]:

$$D^* + A \rightarrow D + A^*$$

The energy transfer from the donor molecule D^* to the acceptor molecule A, which becomes excited, occurs in a non-radiative manner. The efficiency of the process depends on:

– the overlap of the emission spectrum (from the donor) and the absorption spectrum (from the acceptor);

– the distance R_{DA} between the donor and the acceptor. The energy transfer rate k_{dd} can be written as:

$$k_{dd} = \frac{1}{\tau_D^0}\left(\frac{R_0}{R_{DA}}\right)^6 \qquad [3.15]$$

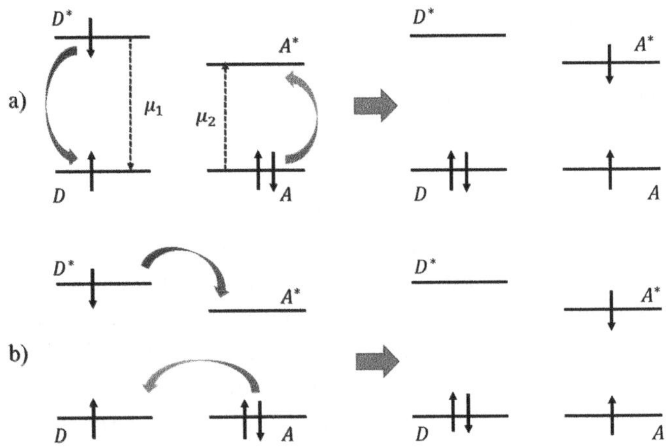

Figure 3.5. *Process of energy transfer between molecules: a) transfer by Förster resonance; b) transfer by Dexter exchange. For a color version of this figure, see www.iste.co.uk/nguyen/electronics1.zip*

Here, τ_D^0 is the lifetime of the donor when no energy transfer occurs, and R_0 is the critical distance or the Förster radius. This value represents the distance between two molecules for which the transfer rate and the decay rate of the excited donor are identical. In other words, at this distance, quenching through an energy transfer is as effective as other processes when no acceptor is present.

In fact, if $R_{DA} = R_0$, then $k_{dd} = \dfrac{1}{\tau_D^0} = k_d$ will be equal to the constant of the emission rate of only the donor. The value R_0 is of the order of a few nanometers compared to the minimum distance between two molecules $(\sim 0.5\,\text{nm})$.

In the Dexter mechanism, transfer is a process of electron exchange between two molecules. The electron of the excited donor molecule D^* is transferred to the acceptor molecule *A*, which becomes excited (state A^*). At the same time, an electron from the fundamental state of the acceptor molecule *A* is transferred to the donor molecule, which is now in the fundamental state (state *D*).

The energy transfer (exchange) rate is written as:

$$k_{ex} = K\, J\, exp\left(-\frac{2R_{DA}}{L}\right) \qquad [3.16]$$

where K is a constant, J is the spectral overlap integral and L is the average Bohr radius of the excited donor and the acceptor at the fundamental state.

3.3.4.2.3. Selection rules

In the energy transfer process, the transitions between states follow the selection rules.

For the Förster mechanism, if the overlap of the emission spectrum (of the donor) and the absorption spectrum (of the acceptor) is effective, the main possible types of energy transfer are:

– energy transfer between singlet states: ${}^1D^* + {}^1A \rightarrow {}^1D + {}^1A^*$;

– energy transfer between triplet and singlet states: ${}^3D^* + {}^1A \rightarrow {}^1D + {}^1A^*$. The phosphorescence of the donor is quenched.

For the Dexter mechanism, the main possible types of energy transfer are:

– energy transfer between singlet states: ${}^1D^* + {}^1A \rightarrow {}^1D + {}^1A^*$;

– energy transfer between triplet states: ${}^3D^* + {}^1A \rightarrow {}^1D + {}^3A^*$. This type of transfer allows for the populating of triplet states of acceptor molecules with a weak or non-existent intersystem conversion;

– energy transfer between triplet and singlet states: ${}^3D^* + {}^3A^* \rightarrow {}^1D + {}^1A^*$ (triplet–triplet annihilation). The donor and acceptor molecules are in the excited state, and the energy exchange through the annihilation of energy allows the two molecules to return to the singlet state, from which fluorescence emissions can be achieved with a delay that is determined by the lifetime of the triplet state, as has already been explained in the process of delayed fluorescence.

3.3.4.3. *Oxidation reaction*

All materials in general, and organic materials in particular, are sensitive to dioxygen and oxidize upon contact. Dioxygen does not absorb light, but becomes active once in contact with a photosensitive material. The fundamental state of dioxygen is a triplet state ${}^{3}O_2$ that becomes excited in the presence of a photosensitizer M to form two excited singlet states ${}^{1}O_2^*$. These two states have different lower energy levels. Oxidation reactions can be written as:

$$ {}^{1}M^* + {}^{3}O_2 \rightarrow {}^{3}M^* + {}^{1}O_2^* \qquad [3.17a]$$

$$ {}^{1}M^* + {}^{3}O_2 \rightarrow {}^{3}M^* + {}^{3}O_2 \qquad [3.17b]$$

$$ {}^{1}M^* + {}^{3}O_2 \rightarrow {}^{1}M + {}^{3}O_2 \qquad [3.17c]$$

$$ {}^{3}M^* + {}^{3}O_2 \rightarrow {}^{1}M + {}^{1}O_2^* \qquad [3.17d]$$

$$ {}^{3}M^* + {}^{3}O_2 \rightarrow {}^{1}M + {}^{3}O_2 \qquad [3.17e]$$

The singlet states of the photosensitive material are inhibited by the reactions *a*, *b* and *c*, while the triplet states ${}^{3}M^*$ are neutralized by the reactions *d* and *e*. The singlet and excited states of dioxygen have high reactivity with the oxidation mechanism and give rise to the quenching of the photoluminescence of many conjugated polymers. For example, the formation of carbonyl groups on the skeleton of the PPV polymer chains constitutes quenchers that decrease the photoluminescence yield of the material (Yan *et al.* 1994). These groups behave like defects that trap charges and suppress radiative transitions between electronic states. In general, structural defects (chain ends and dangling bonds) or chemical defects (impurities) found in the material also lead to the quenching of transitions and affect the intensity of the emission spectra (Gaal *et al.* 2003; Singh *et al.* 2012).

3.3.4.4. *Interchain interactions in conjugated polymers*

When a polymer is diluted in a solution, the excited electronic states are confined to an isolated chain and are *intrachain electronic states*. The transitions between states are not affected by the vicinity of the chain, and the energy efficiency is high. In a film, the chains are in contact or located at very short distances from each other, the energy bands overlap, and the electronic states become *interchain states*. The formation of these species, such as aggregates, affects the electronic transitions and promotes quenching of light emission in the polymer. In conjugated polymers, the aggregates formed may be generated from the overlapping of the segments of a single isolated chain (*intrachain aggregates*) or neighboring chains (*interchain aggregates*) and the conditions for their formation mainly depend on the quality of the solvent, as well as on the temperature and concentration of the polymer.

3.4. Excitons

In inorganic SCs, during the generation process resulting from the absorption of a photon, an electron that has acquired energy higher than the gap energy E_G leaves the valence band and enters the conduction band, becoming free once there. The electron leaving the VB creates a hole in the VB, which is also free to move. If the energy of the absorbed photon is close to but lower than the E_G, the electron is excited towards localized states close to the bottom level E_C of the CB. In this case, the electron is no longer free to move, and remains bound to the hole created in the VB, and the electron–hole pair thus constituted is called an *exciton* and can move in the SC. The average distance between the electron and the hole is called the *Bohr radius* of the exciton. The exciton–electron transitions can be observed through the peaks of the absorption spectrum located near the absorption threshold.

In organic SCs, the absorption of a photon creates electronic transitions and excitation of molecules. When the excited molecular state can be transferred, we speak of an exciton that is composed of an electron and a hole localized in a space and that does not have a dipole moment. That is, the created electrons do not move freely and without limits within the material. On the contrary, they are bound and their movement is restricted. If there is a large electron delocalization together with a large number of molecules, the density of excited states is high, and as a result, the energy levels of singlet and triplet states have energy levels become closer. The Bohr radius of excitons may become comparable to the size of the nanomaterial, and its optical characteristics will change drastically. In general, excitons have a strong reaction to light, and play an important role in the optical spectra of materials.

3.4.1. *Classification of excitons*

There are three types of excitons, which depend on the extent of the electron delocalization (see Figure 3.6).

3.4.1.1. *Frenkel excitons*

Frenkel excitons are electron–hole pairs that are associated with a molecule. The radius of the exciton is comparable to the size of the organic SC molecule or the lattice parameter of the inorganic SC (<0.5 nm). Therefore, these excitons are strongly bound to the molecule and are more or less localized on the basis of the electrostatic interaction of charges.

3.4.1.2. *Wannier–Mott excitons*

Wannier–Mott excitons are electron–hole pairs created in inorganic SCs in which the electron–hole interactions are attenuated by the high permittivity ε of these materials. As a result, the charges are distant, and the Bohr radius of the excitons is within a distance interval of 4–10 nm, that is, several times larger than the size of the molecules. These excitons can potentially delocalize over long distances.

3.4.1.3. *Charge transfer excitons*

These are electron–hole pairs created when a photon is absorbed by adjacent molecules where the hole is on the side of the donor molecule and the electron is on the side of the acceptor molecule.

Intramolecular charge transfers are carried out from the donor molecule to the acceptor molecule. The mean radius of the exciton is between the values of the radii of Frenkel excitons and Wannier–Mott excitons. In other words, these excitons are bound neither too strongly nor too loosely.

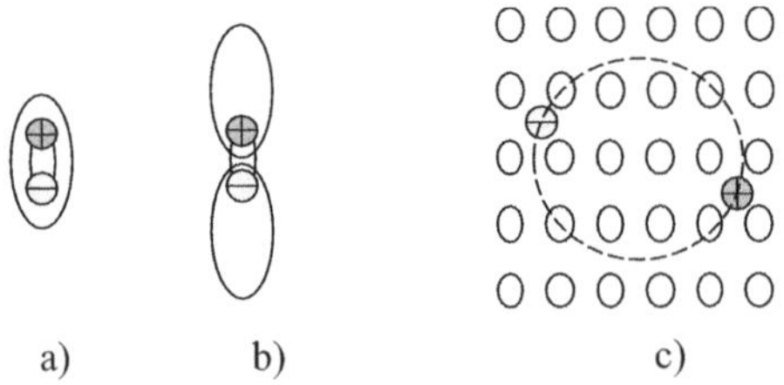

Figure 3.6. *Excitons: a) Frenkel; b) charge transfer; c) Wannier–Mott. For a color version of this figure, see www.iste.co.uk/nguyen/electronics1.zip*

3.4.2. *Binding energy of excitons*

The binding energy of excitons E_B is an important parameter for the study of optoelectronic properties of solids in general and for organic SCs in particular. By definition, this energy is the one necessary to dissociate an exciton in the singlet or triplet state into a pair of free charges.

It can also be defined by the reverse process, that is, E_B is the energy necessary for the recombination of an electron–hole pair for which the relaxation may or may not occur in conjunction with the emission of a photon.

For solar cells, if the energy E_B is low, the separation of charges is made easier and the energy conversion efficiency will be higher. Conversely, for light-emitting diodes, it is of greater interest to have higher energy E_B so as to ensure the recombination of a large number of charge carriers and a better light emission.

In inorganic SCs, the binding energy of the E_B excitons is equal to the energy difference between the SC gap E_G and the energy E_A of the optical absorption threshold:

$$E_B = E_G - E_A \qquad [3.18]$$

The binding energy E_B is low, of the order of approximately 10 meV.

In organic SCs, the difficulty of precisely defining the values of both E_G and E_A at the same time has led to the estimates of very different values of E_B for the same material on the basis of experimental measurements. For example, for poly(phenylene vinylene) (PPV), conflicting results ranging from 0.05 to 1.10 eV have been reported (Leng *et al.* 1994; Marks *et al.* 1994; Singh *et al.* 2012). The reason for this difference is that the values of E_G and E_A depend on the conjugation length of the polymer, and since it is difficult to precisely control this length, different results have been reported.

To define the binding energy E_B, let us consider the different types of excitons (see Figure 3.5) whose energies allow us to consider the concept of binding energy. For a Frenkel exciton, the energy of the exciton E_A is that of the electron–hole pair of a molecule. To form this electron–hole pair from two different molecules of a total energy of E_D, an energy must be provided equal to:

$$E_D - E_A = E_B \quad [3.19]$$

This energy is called the exciton binding energy. According to Knupfer's analysis (Knupfer 2003), we can evaluate E_B using the following method:

– separating the electron–hole pair of the molecule into an electron and a hole that are no longer bound by providing the intramolecular binding energy E_B^{intra};

– placing one of the charges obtained (a hole or an electron) from this molecule into another molecule not adjacent to it by providing the energy U to overcome the repulsion of the molecules whose total charge has been modified. The separation of charges allows for a gain in the kinetic energy equal to the intermolecular energy W separating the two molecules.

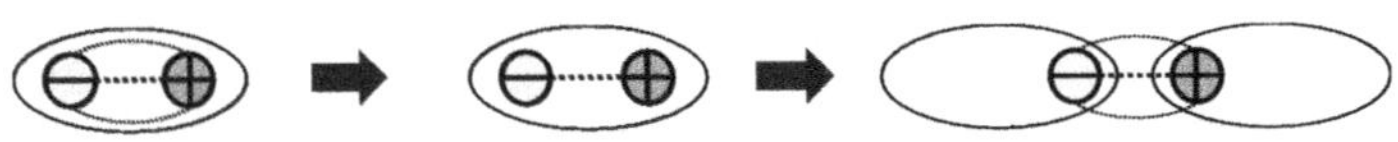

a) b) c)

Figure 3.7. *The steps of the exciton formation: a) initial electron–hole pair; b) separation of charges; c) formation of charge transfer exciton. For a color version of this figure, see www.iste.co.uk/nguyen/electronics1.zip*

Thus, the binding energy E_B is written as:

$$E_B = E_B^{intra} + U - W \quad [3.20]$$

It has been shown that the energies E_B^{intra} and W are approximately equal, and the binding energy of the exciton can be estimated by $E_B \sim U$, that is, the energy of the Coulombic interaction between two adjacent molecules. It is therefore dependent on the distance between the charges of the electron–hole pair, that is, the conjugation length of the polymer. This estimation explains the dispersion in the results when determining the binding energy of excitons, and gives a reasonable value for what this energy will be for a polymer whose conjugation length is about 10 constituent units: ~200 meV.

3.4.3. *Movement of excitons*

In an organic material, the absorption of a photon creates an excited electronic state associated with an exciton. We can consider that the excitation wave covers a domain comparable in size to the wavelength of the absorbed photon. As a result of intermolecular coupling, this domain interacts with the surrounding domains and moves within the material. Thus, the exciton can move in the same way as the domain with which it is associated. We may speak of *exciton diffusion* in a similar way to inorganic SCs: the exciton diffusion is characterized by its *diffusion length* L. This length measures the distance traveled by the exciton during its lifetime τ. The diffusion length is expressed as a function of the *diffusion coefficient* D by the relationship:

$$L = \sqrt{D\tau} \qquad [3.21]$$

Thus, the exciton is not localized on a molecule, but can move over an average distance of about 10 nm, having a lifetime of the order of a few nanoseconds. The average value of the diffusion coefficient is $10^{-4} - 10^{-2}\ \text{cm}^2/\text{s}$ for singlet excitons and $10^{-6} - 10^{-3}\ \text{cm}^2/\text{s}$ for triplet excitons (Ostroverkhova 2016), and the diffusion length L varies from 5 to 20 nm for a singlet exciton and 30 to 300 nm for a triplet exciton. In some organic crystals, the diffusion length may exceed 100 nm for singlets and 100 μm for triplets. Such characteristics are obviously of interest for the absorber materials used in solar cells as they greatly facilitate the charge collection. At the end of this diffusion, the exciton charges separate and combine to emit a photon or turn into heat. This is the consequence of the exciton dissociation process.

3.4.4. *Dissociation of excitons*

Excitons created in the material can be dissociated by providing greater energy than their binding energy E_B, or through traps or an interface that reduces their energy and therefore breaks the exciton binding formed by the electrostatic attraction of the two charges.

In the first case, the Coulombic attraction energy between an electron and a hole, which is separated by a distance r_C, is:

$$E_B = \frac{|e|^2}{4\,\pi\,\varepsilon_r \varepsilon_0\, r_C} = \frac{3\,k\,T}{2} \qquad [3.22]$$

Then:

$$r_C = \frac{|e|^2}{6\,\pi\,\varepsilon_r \varepsilon_0\, kT} \qquad [3.23]$$

Once the charges are separated, if the distance between the electron and the hole is greater than r_C, the thermal agitation is enough to keep them at a distance. The dissociated charges may also reform and recombine. The recombination is called *bimolecular recombination* when the distance between the electron and the hole is greater than r_C and *geminate recombination* when it is less than r_C.

In the second case, defects in the material can trap the electron or the hole of the exciton during the diffusion and dissociate the exciton. This mechanism leads to a reduction in the lifetime of the photoluminescence decay, which is observed experimentally. To improve the mobility of the exciton, it will be necessary to improve the electrical charge transport of the material.

Depending on the process desired for the working of the electronic device, the conditions for the formation, diffusion and dissociation of excitons will be studied, and both the materials and the structure of the device will be chosen to optimize its optical parameters and improve its performance.

3.5. Experimental techniques

Several measurement techniques may be used for the study of photophysical processes. Here, we will examine the most accessible and most commonly used techniques.

3.5.1. *UV–visible absorption spectroscopy*

Absorption spectra allow us to determine the optical characteristics of materials, such as the optical gap, the vibronic transitions of the fundamental state, and the vibronic transitions of the first excited level in the Perrin–Jablonski diagram.

A spectrophotometer includes a light source whose emission covers the UV–visible and near-infrared (NIR) ranges, a dispersive element, and a detector. The incident light beam is divided into a *reference beam* of intensity I_0 and a *measurement beam* of intensity I. The sample absorbs the beam and the interaction with the light results in the attenuation of the wave intensity, which is a function of the wavelength λ. The absorption spectrum represents the absorbance of the material as a function of λ: $A = \log\left(\frac{I_0}{I}\right)$.

When the absorption spectrum is recorded with the sample continuously exposed to the light beam, we speak of *steady-state absorption*, because the corresponding transitions are stable transitions of the permanent regime. To study the transitions of unstable intermediate states with a short or very short duration, a transient absorption spectrophotometer or a pump–probe method is used.

For transient absorption spectroscopy measurements, a laser pulse is used to excite the material for a short time, and then a white light source is used to analyze the change in absorbance as a function of time t. The signal is detected using a transient analyzer with a bandwidth compatible with the width of the pulse. The transient absorbance $\Delta A = \log\left(\frac{I_0}{I_t}\right)$ is thus measured as a function of time t, where ΔA is the difference between the initial absorbance and the absorbance at time t. This method makes it possible to obtain information on the relaxation of the transitions of excited species, and in particular, as a function of the wavelength λ by varying the frequency of the laser.

In the pump–probe method, two signals are used simultaneously to excite the material and detect responses. The pump signal is a laser pulse that excites the material and the probe signal is white light that is delayed with regard to the laser. This delay is used to measure the response of the transient state of absorption. The variation of the absorption as a function of time is $\Delta A = \log\left(\frac{I_{probe}}{I_{pump+probe}}\right)$, which is recorded for different delay times of the probe signal and allows for responses to be analyzed and information to be obtained in the same way as that obtained from transient absorption. This method is primarily used to study transitions of very short durations (<1ns).

3.5.2. *Photoluminescence spectroscopy*

Photoluminescence (PL) spectroscopy is used to study the processes by which materials emit light. There are two different approaches: steady-state PL and time-resolved PL.

3.5.2.1. *Steady-state photoluminescence*

The emission spectrum is obtained by the optical excitation of the material using a light source without taking time into account. The beam is focused on the sample and the output signal is sent to a monochromator and amplified by a photomultiplier. An excitation wavelength λ_{exc} is set for the incident beam, and the emission intensity spectrum (fluorescence or phosphorescence) is recorded as a function of the emission wavelength. It is also possible to record the *excitation spectrum* of fluorescence, that is, the recording of the intensity of the fluorescence emission spectrum at a fixed wavelength λ_{em} when the excitation wavelength λ_{exc} varies. The excitation spectrum allows the identification of transitions responsible for certain particular emissions in composite materials or mixtures of materials.

It should be noted that the intensity of the phosphorescence emission is lower than that of the fluorescence emission, and thus in order to measure it, it is necessary to mask or attenuate the fluorescence spectrum of the material to obtain the adequate intensity and resolution for the spectrum. Furthermore, the phosphorescence spectrum is red-shifted with regard to the fluorescence spectrum; therefore, the spectrum is located in the long wavelength range. To access the phosphorescence spectrum, the recording of the spectrum can be delayed to an instant after the end of the fluorescence emission, and by amplifying the detection of the signal. This prevents the detector from being saturated during the recording.

3.5.2.2. *Time-resolved photoluminescence*

This technique allows for the lifetime of the photo-generated species to be studied by recording the luminescence signal emitted after an excitation of the material of a very short duration (<100 femtoseconds) and with frequencies that are variable from ultraviolet to infrared. In fact, the lifetime of the species varies from a few tenths of a second to a few picoseconds, and the pulses used for the excitation of the material must be shorter than these lifetimes to be able to detect the decay of the response signal. On the other hand, the spectrum must be recorded with an adequate resolution to allow it to separate the components corresponding to different active species. For these spectra, it is necessary to have a time resolution of the order of a few picoseconds. In general, a *streak camera* is capable of meeting the requirements of recording the lifetime and the temporal profile of the measurement.

3.5.2.2.1. Decay equation

To establish the decay equation, let us consider the simplified scheme of the electronic transition processes (see Figure 3.8). Each process is associated with a *rate constant* or *kinetic constant* of the fluorescence or phosphorescence emission.

According to Kasha's rule, the observed fluorescence results from the transitions from the S_1 state. Therefore, the only transitions to be considered are those that occur between S_1 and S_0, regardless of how the original transitions originated, and especially internal conversions and intersystem conversions.

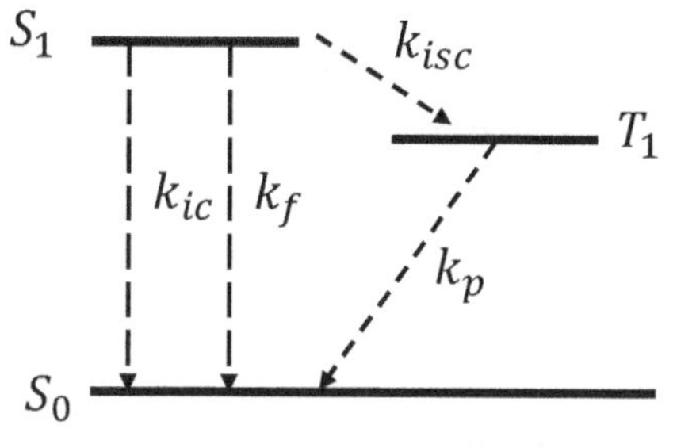

Figure 3.8. *Schematic representation of rate constants of photophysical processes*

Let k_f be the rate constant of fluorescence electronic transitions $S_1 \to S_0$, k_p be the rate constant of phosphorescence electronic transitions $T_1 \to S_0$, k_{ic} be the rate constant of fluorescence internal conversion electronic transitions $S_1 \rightarrow S_0$, and k_{isc} be the rate constant of intersystem crossing electronic transitions $S_1 \rightarrow T_1$.

Considering the nature of photophysical processes, we can simplify the equations by expressing them as $k_r = k_f + k_p$ and $k_{nr} = k_{ic} + k_{isc}$. The first term is the rate constant for *radiative decay* and the second is the rate constant for *non-radiative decay*. The change in the number of excited singlet states *N* as a function of time is expressed by the equation:

$$\frac{dN}{dt} = -\left(k_r + k_{nr}\right)N \qquad [3.24]$$

By integrating this equation with the initial condition $N(t=0) = N_0$, we obtain:

$$N(t) = N_0 \exp\left[-\left(k_r + k_{nr}\right)t\right] \qquad [3.25]$$

The intensity of the fluorescence spectrum is equal to the number of moles of photons emitted per second and per unit of volume, which is written as:

$$I_F = k_F \times N(t) = k_F N_0 \exp\left[-\left(k_r + k_{nr}\right)t\right] \qquad [3.26]$$

The *fluorescence lifetime* is defined as the time constant of the exponential decay of the signal by:

$$\tau_f = \frac{1}{k_r + k_{nr}} \qquad [3.27]$$

The fraction of radiative transitions is therefore $\dfrac{k_r}{k_r + k_{nr}}$, and thus the fraction of non-radiative transitions is $\dfrac{k_{nr}}{k_r + k_{nr}}$.

In the steady-state emission regime, the fluorescence quantum yield Φ_f is defined as the ratio of the number of photons emitted to the number of photons absorbed. Thus, we can write:

$$\Phi_f = \frac{k_r}{k_r + k_{nr}} = \frac{\tau_f}{\tau_r} \qquad [3.28]$$

where $\tau_r = \frac{1}{k_r}$ is the lifetime of the radiative transitions. Having established the quantum yield and the lifetime of the fluorescence, we can deduce the radiative lifetime τ_r.

3.5.2.2.2. Multiple decay process

We can observe that, in many cases, exciton decays in organic materials are multiple processes; in other words, they are electronic transitions of a different nature. The experimental decline curve is not a single exponential curve, but instead has two or more components (multiexponential kinetics). Now, we will examine the case of two independent transition processes and write out equation [3.24] for each of these processes:

$$\frac{dn_1}{dt} = G(t) - k_{r1} n_1 \qquad [3.29]$$

$$\frac{dn_2}{dt} = G(t) - k_{r2} n_2 \qquad [3.30]$$

where n_1 and n_2 are the populations of excited singlet states, k_{r1} and k_{r2} are the decay rates for each species, and $G(t)$ is a function related to the response of the measurement apparatus. The total population of excited states is $n = A_1 n_1 + A_2 n_2$, where A_1 and A_2 are constants proportional to the intensity of the photoluminescence signal. In this case, there are two exponential functions that are superimposed. The adjustment of the experimental decay curve will provide the values of the decay rates that will be used to interpret the transition processes. In general, a process with short times and rapid decay k_{r1} indicates a diffusion of excitons in the material and a process with long times and a longer decay rate k_{r2} indicates a thermalization of excitons.

3.5.3. *Infrared and Raman spectroscopy*

3.5.3.1. *Infrared spectroscopy*

In the infrared domain, where wavelengths are greater than 1 micron, the electronic transitions correspond to molecular vibrations. A molecule absorbs radiation if its vibrational frequency ν_v is the same as that of the radiation ν : $\nu = \nu_v$ (see Figure 3.9).

For a molecular vibration, the movements of the atoms can induce a change of the dipolar moment, and these vibrations are said to be *infrared-active*. Other vibrations that do not generate a variation of the dipole moment are said to be *inactive* because there is no absorption or emission of light.

In an infrared absorption spectrum, vibrational bands of high intensity and others of low intensity can be observed. The fundamental bands correspond to the transitions from the fundamental state to the first excited state, and such a transition is called *the fundamental transition*. The other bands result from transitions from the fundamental state to the higher excitation states, and are called *harmonics*. Couplings may be produced between vibration modes that change the energy and therefore the position of the absorption peaks. Two frequency and vibration modes, ν_1 and ν_2, can generate vibrations of frequencies $(\nu_1 + \nu_2)$ and $(\nu_1 - \nu_2)$ by association.

An analysis of absorption spectra in the infrared domain provides information on functional groups and allows for a quantitative characterization of the organic materials. For the study of polymers, spectral domains of particular interest include the region between 4,000 and 3,000 cm^{-1} where the absorption bands of the C–H and O–H bonds are located. This region is called the *aromatic region*. The absorption band at approximately 1,600 cm^{-1} is studied for the formation of carbonyl bonds, which can indicate the possible oxidation of the sample surface when they appear and grow.

The absorption spectrum is customarily represented as a function of the wavenumber $\bar{\nu}$ or the inverse of the wavelength λ:

$$\frac{1}{\lambda} = \bar{\nu} = \frac{\nu}{c} \qquad [3.31]$$

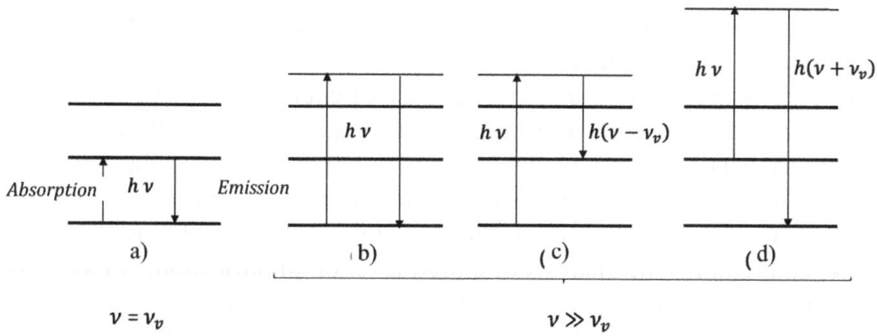

Figure 3.9. *Transitions in the processes of: a) infrared absorption; B) Rayleigh scattering; c) Stokes Raman scattering; d) anti-Stokes Raman scattering*

3.5.3.2. *Raman spectroscopy*

In addition to infrared absorption, Raman spectroscopy is often used to study the vibrational transitions of organic materials. The principle of this technique of analysis is based on the scattering of the photons of a laser of frequency ν_0 that radiates the material. When many of the photons are scattered with the same frequency ν_0 as the incident radiation, the process is called Rayleigh scattering. When a small portion of the photons are scattered with a modified frequency ν_d, the process is called *Raman scattering* (see Figure 3.9).

For $\nu_d < \nu_0$, the gap in the scatter frequency is $\nu_v = \nu_0 - \nu_d$ and $\nu_d = \nu_0 - \nu_v$, and this scatttering is called *Stokes Raman scattering.*

For $\nu_d > \nu_0$, $\nu_d = \nu_0 + \nu_v$, the scattering is called *anti-Stokes Raman scattering.*

There is a rule known as the mutual exclusion rule, which states that if the molecule has a center of symmetry, the symmetrical vibrations with respect to this center are active in Raman scattering and inactive in infrared absorption. Conversely, antisymmetric vibrations are active in infrared absorption and inactive in Raman scattering.

The Raman spectrum generally presents a wavenumber shift $\bar{v}_v$ between the Rayleigh line and the Stokes Raman lines (Stokes and anti-Stokes). This difference is independent of the frequency of excitation radiation, which is characteristic of electronic transitions of the material. The spectrum includes vibrational bands that provide information on vibration modes, the symmetry of the mode, the concentration of the scattering molecules and the interactions of the molecules. In general, the intensity of Raman bands is weak, and requires the analysis of a high level of amplification. A technique is used to enhance certain bands of the spectrum using the resonance of the vibrations of the chromophores involved in the transitions. This technique, called *resonance Raman scattering*, uses a wavelength of excitation radiation close to that of an electronic transition of the material. The resonance specifically amplifies the corresponding Raman scattered radiation and allows the transition to be accurately analyzed. The *surface-enhanced Raman scattering* (SERS) effect takes advantage of the resonance of the molecules adsorbed by a nanostructured metal surface to study the morphology and processes of the surface. The intensity of the Raman response of the molecules in the vicinity of the surface is amplified through the amplification of the electromagnetic field scattered by metal particles. As a result, the SERS technique allows for the detection of molecules with low concentrations on the surface of the material in contact with a metal film by generating Raman spectra of a significant intensity.

In the study of polymers and organic materials, Raman spectroscopy is widely applied in measuring physical characteristics. One example of an application of Raman spectroscopy is to determine the number of single layers of graphene (Malard *et al.* 2009). The Raman spectrum of graphene generally consists of two bands: band *D* located at $\sim 1,285\,\text{cm}^{-1}$, corresponding to the defects, and the characteristic band *G* located at $\sim 1,582\,\text{cm}^{-1}$, corresponding to the vibrations of sp2 carbon atoms. A third band *G'* or *2D* localized towards $\sim 2,700\,\text{cm}^{-1}$ can allow for an estimation of the number of graphene layers based on the changes in the shape, intensity and position of the maximum band of the spectrum of the monolayer. Another example is the determination of the conjugation length of PPV by the analysis of the Raman spectra of phenylene-vinyl oligomers (Nguyen *et al.* 1999). A study has been conducted on the change in the scattered frequency of the triplet of the Raman band located at $\sim 1,600\,\text{cm}^{-1}$ in phenylene-vinylene

(PV) oligomers with a known number n of repeat units. This band is characteristic of the vibrations of phenyl and vinyl groups of polymer chains, and the analysis of the frequency shift of this band in oligomers allows for the effective conjugation length of the PPV to be estimated, which will be of the order of ten times the repeat unit length.

4

Electronic Processes

4.1. Introduction

The operation of an organic electronic device requires electrical connections between the electrodes of the device and a power supply and/or measurement apparatus. When a voltage is applied to the device, the charge carriers are injected into the active material via the electrodes and move to the output electrode. Figure 4.1 provides a schematic description of the electrical transport of the charge carriers, holes and electrons, and the different processes that take place in the materials that make up the device, presented here with a simplified structure.

Under the effect of the applied electric field, the charge carriers are injected via the electrodes into the organic material, electrons at the cathode and holes at the anode (process 1, *injection*). The injection takes place through a potential barrier created by the contact between the metal and the SC. The charge carriers then move throughout the width of the active layer (process 2, *transport*) following different conduction mechanisms that are dependent on the structure of the material. During their transport, carriers can interact with various charged species in the material (processes 3, *interactions*) which can result in different consequences (temporary disappearance by trapping, definitive disappearance by recombination, etc.). Overall, the quality of the transport is estimated by the *mobility*, which expresses the ability of carriers to move within the material. Finally, once they have arrived at the opposite electrode, the carriers are evacuated into the external circuit through the contact between the SC and the metal. The current measured at the output is called the *leakage current* of holes (at the cathode) or electrons (at the anode).

In this chapter, we will study the different stages of transport, starting with the basis of the knowledge of electrical processes in inorganic electronics that we presented in Chapter 1, and considering the unique characteristics of the electronic structure of organic materials.

In fact, the amorphous nature of the organic material is related to a structural disorder, and creates many localized states in the gap of the SC. As a result, the charges move by hopping between states. On the other hand, the interactions between molecules are low compared to the covalent bonds of inorganic SCs. As a result, the width of the transport bands is small (10– 100 meV) and the probability of charge transfer between molecules is low, leading to reduced mobility of the carriers. Furthermore, the relative permittivity of most organic materials is low ($\varepsilon_r \sim 3$), which leads to high values for the binding energy of the excitons.

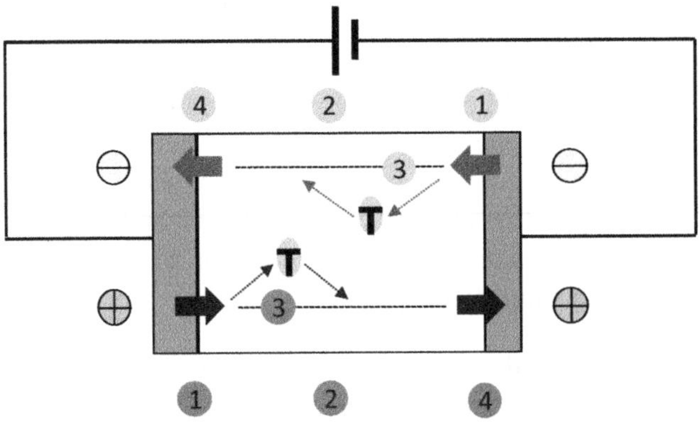

Figure 4.1. *Schematic representation of the movement of charge carriers in an organic electronic device: 1) injection through the metal/SC contact, 2) transport, 3) interactions with other charged species inside the material, 4) extraction through the SC/metal contact. For a color version of this figure, see www.iste.co.uk/nguyen/electronics1.zip*

4.2. Charge carrier injection process

To better illustrate this process, Figure 4.2 shows the energy diagram of the basic structure of an organic light-emitting diode that has an organic SC thin film and two electrodes. In a simple structure, we can distinguish the two interfaces formed at the cathode and anode, respectively. On the cathode, the interface is formed by depositing a metal film M onto the surface of the SC, and on the anode, the interface is formed by depositing the organic film onto a transparent conducting film.

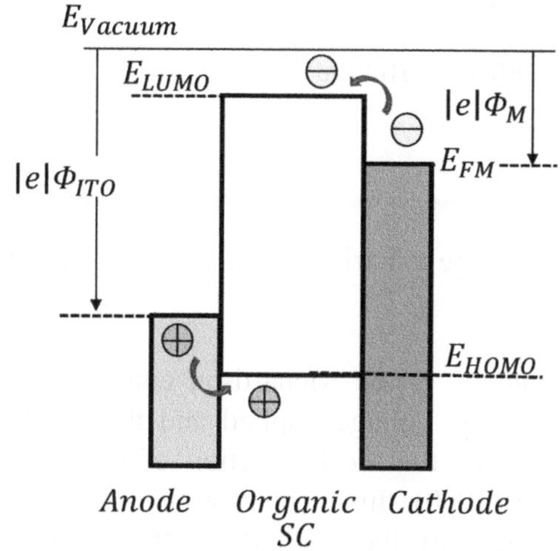

Figure 4.2. *Injection process of charge carriers in a simple diode structure. For a color version of this figure, see www.iste.co.uk/nguyen/electronics1.zip*

4.2.1. *Injection mechanisms*

There are two possible mechanisms by which electrons can be injected into an organic SC.

4.2.1.1. *Thermoionic emission or Schottky effect*

The first mechanism is the thermoionic emission of electrons above the Schottky barrier formed between the SC and the metal.

This process occurs if the electron from the metal acquires enough thermal energy to overcome the potential barrier $|e|\Phi_{bn}$, for which a decrease $\Delta\Phi$ will depend on the image potential and the applied voltage V_0. The current density is written as:

$$J_{sc} = J_s \left[exp\left(\frac{|e|\,V_0}{k\,T}\right) - 1\right]$$

$$J_{sc} = \left[A^* T^2\, exp\left(-\frac{|e|\Phi_{bn}}{k\,T}\right) exp\left(\frac{|e|\Delta\Phi}{k\,T}\right)\right]\left[exp\left(\frac{|e|\,V_0}{k\,T}\right) - 1\right] \qquad [4.1]$$

This expression can be written in the following form:

$$J_{SC} = A^{*}T^{2}\, exp\left[-\,\frac{|e|\,\Phi_{bn} - \beta_S\, E^{1/2}}{k\,T}\right] \qquad [4.2]$$

where β_S is the Schottky coefficient:

$$\beta_S = \sqrt{\frac{|e|^3}{4\,\pi\,\varepsilon_0\,\varepsilon_r}} \qquad [4.3]$$

and $E = \frac{V_0}{d}$ is the average electric field in the organic film, where d is the thickness of the film.

In this expression, A^* is the Richardson constant. The current density therefore depends on the voltage applied and the height of the metal/SC potential barrier, that is, the work function of the metal of the cathode. According to Bässler (2000), this mechanism is unlikely to occur, because a change in the intensity of the applied electric field E from 10^5 to 2×10^6 V cm^{-1} would result in a lowering of the Schottky barrier from 0.06 to 0.28 eV for a polymer of a relative permittivity of $\varepsilon_r = 3.5$. In other words, the reduction that occurs is considerable, reaching a level that is nearly that of the height of the barrier itself.

4.2.1.2. *Fowler–Nordheim effect*

The second possible mechanism is the Fowler–Nordheim effect, by which electrons subjected to an electric field are capable of crossing a potential barrier when it is sufficiently thin. The metal/SC contact forms a triangle-shaped potential barrier in which the vertex corresponds to the maximum of the potential. When an electron receives sufficient energy, it rises, approaches the top of the barrier, and is able to cross it by tunneling. The current density predicted by this theory is a function of the applied electric field E as given by:

$$J_{FN} = B\,E^{2}\, exp\left(-\,\frac{\kappa}{E}\right) \qquad [4.4]$$

In this expression, B is a constant and the numerator κ of the exponent is given by:

$$\kappa = \frac{8\pi\,\sqrt{2\,m_n^*}\,{\Phi_{bn}}^{3/2}}{3\,|e|\,h} \qquad [4.5]$$

where m_n^* is the effective mass of the electron and h is Planck's constant.

According to the expression [4.4], at a constant temperature, the curve $Ln\left(\frac{J_{FN}}{E^2}\right)$ as a function of $\frac{1}{E}$ is a straight line whose slope determines the potential barrier Φ_{bn} at the interface of the SC and the metal.

By varying the height of the potential barrier Φ_{bn} by changing the work function of the electrode metal, we can experimentally verify the charge injection mechanism into the cathode. We can also apply this same method to the injection of holes on the anode side.

Parker (1994) has studied the injection of charges into the cathode of MEH-PPV-based diodes in detail, and has reached conclusions in favor of the Fowler–Nordheim effect. Other investigations of conjugated polymer diodes have reached different conclusions, and there is no consensus on the mechanism for the charge injection at the metal/SC interface. The process seems to depend on the history of the sample studied (the polymer used, shaping method, polymer/metal interactions, etc.).

4.2.2. *Hole or electron devices*

In inorganic SCs, the process of doping with impurities allows N-type or P-type SCs to be obtained with controlled concentrations. This cannot be done with organic SCs. As we have seen earlier, doping is possible and can be carried out with these materials. However, in the majority of cases, the organic material becomes unstable after doping, and in practice, doping is rarely used to create SCs of a determined type for the production of organic electronic devices. In these devices, the charge carriers are injected from the electrodes into the organic material though processes occurring at the interface that can be summarized as follows. At the cathode, the SC/metal contact is a Schottky contact and the electrons can be injected into the SC under the effect of an applied electric field. At the anode, the potential barrier for the holes is equal to the difference in the ionization energy of the transparent conducting oxide (usually ITO) and the HOMO level of the organic material. With the electric field applied, the holes that have acquired sufficient energy flow over the potential barrier and are thus injected into the HOMO band of the SC.

The injection processes are specific to organic devices because the energy bands of the materials suitably correlate with those of electrode metals and ITO. On the anode side, despite the existence of other TCO materials, ITO is currently the only oxide in practical use in the anodes of the devices that are brought to market. On the cathode side, the choice of metal will focus on those that form low-potential barriers to facilitate the injection of electrons. These metals must have a low work function to minimize the height of the barrier. In practice, the electron affinity of the most commonly used conjugated polymers varies between 2.5 and 3 eV and the work function of the metals used as cathodes varies from 2.8 to 5.5 eV. The potential barriers for electrons can vary from 0.3 to 3 eV. As an example, Table 4.1 gives the values of the work function of metals and the height of the potential barrier for electrons in the case of an MEHPPV/metal contact. Depending on which metal is chosen, the potential barrier can be either very low or very high.

Metal	**Ca**	**Mg**	**Ag**	**Al**	**Au**
Work function (eV)	2.9	3.7	4.2	4.2	5.1
Potential barrier (eV)	0.1	0.9	1.4	1.4	2.3

Table 4.1. *Metal work function and potential barrier for an MEHPPV/metal contact*

Theoretically, the injection of electrons into the cathode is facilitated by the use of calcium, barium or magnesium electrodes. It becomes more difficult with metals such as aluminum or gold. Experimentally, very few results have been obtained that are in line with the projections made for the measurements of the current–voltage characteristics. In these experiments, a comparison was made of the I(V) characteristics of diodes with an identical geometric structure and dimensions, with the exception of the cathode metal. It has been determined that the current density in diodes using metals with a low work function is higher than that of diodes with a high work function. However, the changes in the current density are not in accordance with the value of the potential barrier in Table 4.1. The reason for this divergence will be discussed in the next chapter, which addresses interface processes. Nevertheless, it is useful to consider this characteristic that allows the creation of structures with majority charge carriers of a defined type (N or P). As we have noted, it is not practical to use doping in organic materials to obtain a given type of conduction, as is done in inorganic SCs. However, we can create majority carriers of the predetermined type in an organic diode by preventing the injection of carriers of the other type at the level of the injection electrode. For example, to create N-type charge carriers, that is, electrons in the diode, the holes normally injected from the anode in the SC have to be blocked. To do this, it will suffice to use materials with a much lower work function than the energy of the HOMO band to create potential barriers high enough for the anode, preventing the holes from entering the diode. Such devices are called *electron-only devices* (see Figure 4.3). *Hole-only devices* are designed in the same way, using metals with high levels of work function at the cathode, thus making it difficult to inject electrons into the material. These types of devices are of particular interest for the study of electrical transport because they are equivalent to doped inorganic SC devices.

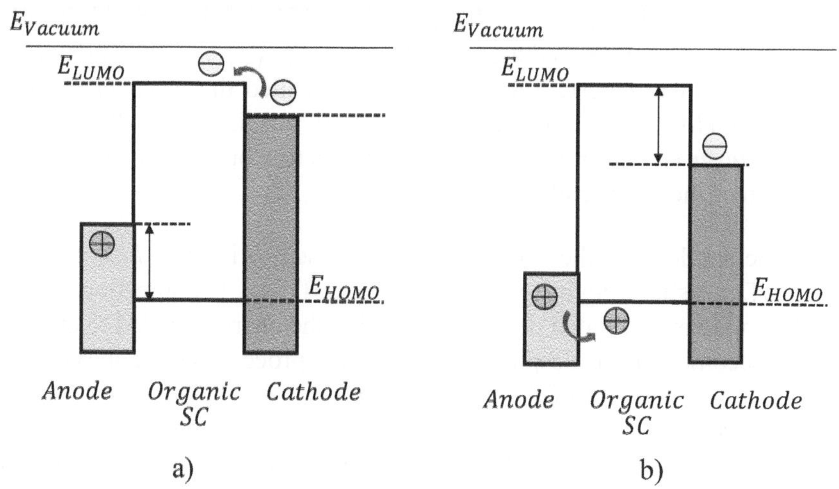

Figure 4.3. *Energy band diagram of electronic devices: a) electron-only; b) hole-only. For a color version of this figure, see www.iste.co.uk/nguyen/electronics1.zip*

4.2.3. *Transport layers*

The injection of electrons into the cathode of an organic diode depends on the potential barrier at the organic/metal SC contact, that is, the work function of the metal when the organic material has been selected. To minimize the potential barrier, metals such as calcium, barium and magnesium are selected for the cathode. The work function for these is close to that of most conjugated polymers and allows a high injection of electrons, and therefore a high current density. However, these metals are chemically unstable and oxidize easily, which changes the structure of the original interface, and the barrier quickly becomes an insulator, preventing the flow of electrons. High-work-function metals are generally chemically stable, but their potential barrier is high, which makes the injection difficult. To overcome this disadvantage, the height of the barrier is changed by introducing an intermediate energy level E_{LUMOi} between the LUMO level of the organic material and the Fermi level E_{FM} of the metal. This level modifies the electron injection process in the organic SC in two steps:

– step 1: $E_{FM} \rightarrow E_{LUMOi}$;

– step 2: $E_{LUMOi} \rightarrow E_{LUMO}$.

The potential barriers that must be overcome by electrons are lower than those that originally exist, and the energy needed to inject the electrons is also smaller. The intermediate energy level E_{LUMOi} corresponds to the LUMO level of a material, generally organic, called the *electron transport material* (ETM). The thin film of this material is called the *electron transport layer* (ETL) and is deposited on the active organic layer before the deposition of the metal cathode.

On the anode side, the same technique is used to insert a *hole transport layer* (HTL), using a *hole transport material* (HTM). The injection of holes into the anode is facilitated due to the fact that the energy needed to cross the barriers becomes lower, despite the fact that the process is done in two steps. In addition to the effect of the height of the potential barrier being reduced, the transport layers act as buffers between the electrode and the active layer and prevent direct contact between the two materials, which can generate interactions and create sources of defects or degradation.

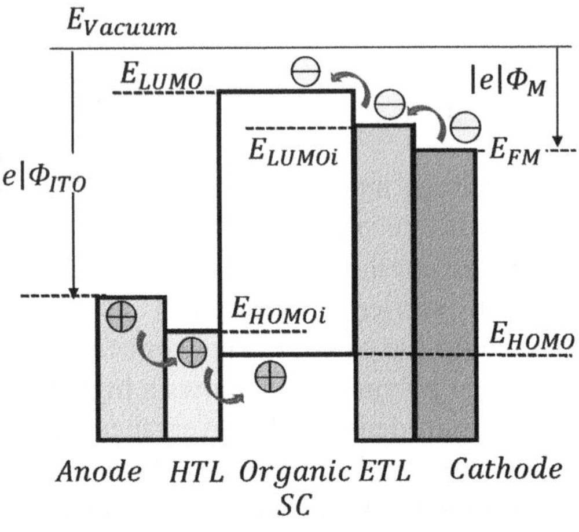

Figure 4.4. *Energy band diagram of the electronic device with transport layers of electrons (ETL) and holes (HTL). For a color version of this figure, see www.iste.co.uk/nguyen/electronics1.zip*

4.3. Charge transport process

Once injected into the organic SC, the charge carriers are transported through the effects of the electric field $\vec{E}$: the holes move in the same direction as the applied field, and the electrons move in the opposite direction. These movements can be characterized by the charge mobility μ, which describes the ability of the carriers to move in the material.

During their movement, the carriers may interact with different species on their trajectory (see process 3 in Figure 4.1) and their mobility is affected by these interactions. Defects in the material, which capture the carriers and prevent them from moving, are the main factor in reducing mobility.

We can distinguish two types of charge movement in a polymer: movement along a polymer chain, called *intra-chain conduction*, and movement between chains, called *inter-chain conduction* (see Figure 4.5). For small molecules, no chains are formed, and conduction occurs between molecules (*inter-molecule conduction)*.

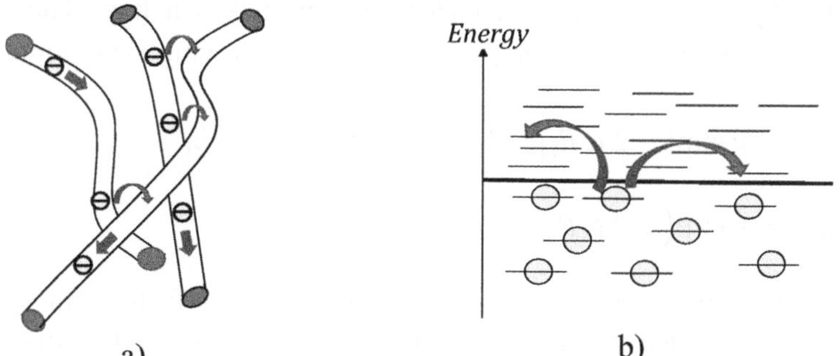

Figure 4.5. *a) Intra-chain and inter-chain movements of electrons; b) energy diagram of electron hops between two sites. For a color version of this figure, see www.iste.co.uk/nguyen/electronics1.zip*

In intra-chain conduction, the movement of the carriers occurs along the delocalized molecular orbitals. The direction of the movement is that of the electric field, and the orientation depends on the sign of the charge. The orientation of the polymer chains is generally random; some charge carriers will hop between the chains in the direction of the applied field when allowed by the configuration (see Figure 4.5a).

4.3.1. *Hopping mechanisms*

In this process, the transport of electrons in the material consists of a transfer of charges between two adjacent molecules or between two segments of a polymer chain that are considered as two nearby energy sites.

4.3.1.1. *Miller–Abrahams model*

The Miller–Abrahams model (Miller and Abrahams 1960) was formulated for the transport of charge carriers in doped inorganic SCs. It is based on the effects of doping of SCs on transport at low temperatures where the conduction occurs between the impurity atoms. When the concentration of impurities is low, the hopping mechanism of carriers assisted by phonons is done through the transition of carriers from an occupied state located below the Fermi level E_F to an unoccupied state located above E_F. This process allows the movement of charges in these materials under such conditions.

Consider two sites of energy levels E_i and E_j, separated by a distance equal to R_{ij}, between which a charge carrier makes a hop. The rate of transitions κ_{ij} from site i to site j is given by the following expression:

$$\kappa_{ij} = \nu \exp\left(-2\,\alpha\, R_{ij}\right) \exp\left(-\frac{E_j - E_i}{k\,T}\right) \text{ if } E_j > E_i \qquad [4.6a]$$

$$\kappa_{ij} = \nu \exp\left(-2\,\alpha\, R_{ij}\right) \text{ if } E_j < E_i \qquad [4.6b]$$

In these expressions, ν is the hopping frequency factor and α is the *overlap factor*, which is equal to the inverse of the delocalization distance d. The first exponential term expresses the probability that a transfer will occur between sites, which depends on the distance between them. The second exponential term represents Boltzmann's probability that a hop will occur from a lower-energy site to a higher-energy site. This factor has the value of 1 for a hop from a higher-energy site to a lower-energy site. When an electric field $\vec{E}$ is applied, the numerator of the exponent of the second term is multiplied by $(-|e|\,\vec{r}.\vec{E})$, where $\vec{r}$ is the vector connecting the centers of the two sites. The applied field facilitates the hop upward but has no effect on a downward hop under this model.

The current due to the carrier hopping is calculated using transition rates and the probability of the occupation of the sites i and j.

4.3.1.2. *Mott–Davis model*

This model is known as the *variable-range hopping* (VRH) model (Mott 1993). It is based on the assumption that at low temperatures, the hops the carriers make between localized sites do not necessarily occur between their nearest neighbors, but instead are more likely to occur in such a way as to use a minimum of energy. Consider a site occupied near the Fermi level E_F and a domain surrounding this site of radius R. The number of states per unit of energy of this domain is equal to:

$$\frac{4\,\pi}{3}\,R^3\,N(E_F)$$

where $N(E_F)$ is the density of states at the Fermi level (by eV and by unit of volume).

For a hop between two distant sites of R to be made with a minimum of energy, the amount of energy consumed will be equal to:

$$\Delta E = \left[\frac{4\,\pi}{3}\,R^3\,N(E_F)\right]^{-1}$$

The probability of process transitions P takes into account the probability of a hop over the distance R separating the two sites and Boltzmann's probability of hops between two sites with different energy levels of ΔE.

$$P \propto \exp(-\,2\,\alpha\,R) \times \exp\left(-\frac{\Delta E}{k\,T}\right) \qquad [4.7]$$

This expression is similar to expression [4.6a], which was inspired by the Mott study. The probability that hops will occur as a function of the distance is maximum if:

$$\frac{dP}{dR} = 0$$

The sum of the terms:

$$2\,\alpha\,R + \frac{\Delta E}{k\,T} = 2\,\alpha\,R + \frac{1}{\frac{4\,\pi}{3}\,R^3\,N(E_F)\,k\,T} \qquad [4.8]$$

is therefore minimal. We obtain:

$$R = \left[\frac{9}{8\,\pi\,N(E_F)\,\alpha\,kT}\right]^{1/4} \quad [4.9]$$

By replacing the value of R in expression [4.7], we obtain the probability of hops P and, as a result, the electrical conductivity σ in the following form (*Mott's law*):

$$\sigma(T) = A\,exp\left(-\frac{B}{T^{1/4}}\right) \quad [4.10]$$

where:

$$B = \frac{4\sqrt{3}}{3}\left(\frac{2}{\pi}\right)^{1/4}\left[\frac{\alpha^3}{k\,N\,(E_F)}\right]^{1/4} \sim 2{,}063\left[\frac{\alpha^3}{k\,N\,(E_F)}\right]^{1/4} \quad [4.11]$$

Experimental studies of the conductivity as a function of temperature of doped or amorphous crystalline semiconductors give an expression similar to that of [4.10] in the form of:

$$\sigma(T) = A\,exp\left(-\frac{B}{T^{n}}\right) \quad [4.12]$$

The exponent n in this expression is $n = \frac{1}{4}$ but will take the value of $n = \frac{1}{2}$ when there is a strong Coulombic interaction of electrons in the material. However, it is difficult to distinguish the mechanisms by which the hops occur on the basis of the values of the exponent, because the experimental curves $Ln\,\sigma\,(T^{-n})$ when $n = \frac{1}{2}$ and $n = \frac{1}{4}$ are both straight lines very close to each other. On the other hand, in organic SCs, the hopping distances determined experimentally are often high, greater than the dimensions of the molecules, which makes it difficult to interpret the measurements.

4.3.1.3. *Charge carrier mobility*

In inorganic SCs, the mobility μ of the charge carriers is high and varies between 10^3 and 10^5 cm^2 V^{-1} s^{-1}. For electrons, it is expressed using the electron current density as:

$$J = |e|\,n\,\langle v\rangle = |e|\,n\,\mu_n\,E \quad [4.13]$$

where n is the concentration of carriers, $\langle v \rangle$ is the average velocity of carriers (electrons) and E is the electric field.

In organic SCs, the mobility of carriers is lower in comparison with that of inorganic SCs. It varies experimentally between 10^{-6} and 10^{1} $cm^2 V^{-1} s^{-1}$ and depends on the concentration of the carriers and the electric field. However, there is no consensus on the quantitative dependence of the mobility of n and of E, as there are several theories that have been put forward to explain the results of the mobility measurements.

For the vast majority of organic materials, the dependence of the mobility of the electric field and the temperature can be described by the equation:

$$\mu\,(E,T) = \mu_0(T)\,exp(\beta\,\sqrt{E}) \qquad [4.14]$$

with:

$$\mu_0(T) = \mu_\infty\,exp\left(-\frac{\Delta E}{k\,T}\right) \qquad [4.15]$$

and:

$$\beta = B\left(\frac{1}{k\,T} - \frac{1}{k\,T_0}\right) \qquad [4.16]$$

In the above expressions, the parameters $\mu_\infty, \Delta E, B$ and T_0 are characteristic of the organic material. This equation indicates that the mobility of the carriers is thermally activated and is a function of the square root of the applied electric field.

This form of dependence is similar to that of the Poole–Frenkel effect, which governs the transport in the bulk of insulating materials and is called the *Poole–Frenkel mobility*.

On the basis of the model of the carrier hops, Bässler (1993) proposed a model called the *Gaussian disorder model* (GDM) as a way to evaluate mobility. According to this model there is a Gaussian distribution of the density of states of energy $g(E)$ in which the carriers move by hopping. The distribution is defined by (see Figure 4.6):

$$g(E) = \frac{N_{sites}}{(2\,\pi\,\sigma_D)^2} \times exp\left[-\frac{(E-E_0)^2}{2\,\sigma_D{}^2}\right] \qquad [4.17]$$

where N_{sites} is the density of the distribution sites, σ_D is the standard deviation or width of the distribution and E_0 is the energy of the center of distribution.

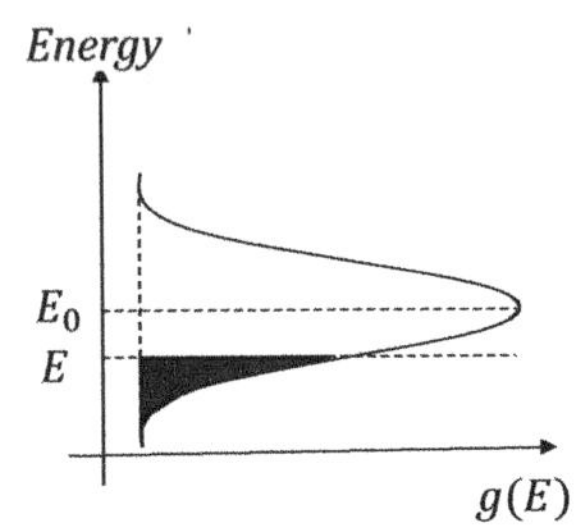

Figure 4.6. *Density of states following the Gaussian disorder model (GDM). For a color version of this figure, see www.iste.co.uk/nguyen/electronics1.zip*

This function allows the introduction of the notion of *transport site disorder* in the *density of states* (DOS) due to the deformation of the segments of chains and the loss of their conjugation length. There are two types of disorder to consider. The first of these, *energetic disorder*, is due to the variations in the inter-site distances, which generate variations in the distribution of energy levels. This disorder is characterized by the parameter σ. The second is the *off-diagonal static disorder*, characterized by the positional disorder parameter Σ. This disorder occurs due to the coupling of molecules with different orientations.

The mobility of carriers under these conditions is expressed by:

$$\mu_{GDM} = \mu_0 \times exp\left[-\left(\frac{2\,\sigma}{3\,k\,T}\right)^2\right] \times exp\left\{C\left[\left(\frac{\sigma}{k\,T}\right)^2 - \Sigma^2\right]\sqrt{E}\right\} \qquad [4.18a]$$

$$\mu_{GDM} = \mu_0 \times exp\left[-\left(\frac{2\,\sigma}{3\,k\,T}\right)^2\right] \times exp\left\{C\left[\left(\frac{\sigma}{k\,T}\right)^2 - 2{,}25\right]\sqrt{E}\right\} \qquad [4.18b]$$

where μ_0 is the mobility limit when $T \rightarrow \infty$ and C is a constant that depends on the distance separating two sites. Equation [4.18a] is valid for $\Sigma \geq 1.5$ and equation [4.18b] is valid for $\Sigma < 1.5$.

The mobility calculated by this model is similar to the expression of Poole–Frenkel mobility and allows us to relate the strong dependence of the temperature of $\mu(T)$ in organic SCs to the processes of carrier-hopping in a medium which is energetically disordered. Many other models have been proposed to improve the GDM model, which remains the first that has been established to explain the dependence $\mu(E, T)$ in organic SCs.

4.3.1.4. *Measurements of charge carrier mobility*

The mobility of charge carriers is a key parameter for studying the charge transport processes in SCs. The Hall effect is used to measure the mobility of electrons or holes in inorganic SCs. The high values of the carriers allow high-quality measuring devices to be used to obtain results with good accuracy. This technique is difficult to adapt to organic SCs because the mobility of the carriers is very low and other techniques should thus be used for the measurement of mobility in these organic materials. Here we present two commonly used measurement techniques, keeping in mind that there are many other possible ways to determine the mobility of organic SC carriers.

4.3.1.4.1. Time-of-flight technique

The principle of the *time-of-flight* (TOF) technique is illustrated in Figure 4.7. It consists of measuring the transit duration (or time) τ_T of the charge carriers created near a face of the sample of thickness d and illuminated by a light excitation. Under the effect of an electric field $\vec{E}$ applied to the sample, the charge carriers move and are detected at the output of the sample by an oscilloscope, for example. The speed v and the mobility μ of the carriers are determined by the expression:

$$v = \frac{d}{\tau_T} = \mu E \quad [4.19]$$

In the example of the setup of Figure 4.7 for measuring electron mobility, we consider an electron–hole pair created by the excitation of light at the cathode. An electric field is applied, separating the charges: the hole is removed by the cathode and the electron moves to the anode. The electrons create a constant current $I(t)$ that ends once they reach the anode, and the transit time represents the time taken by the electrons to travel this distance.

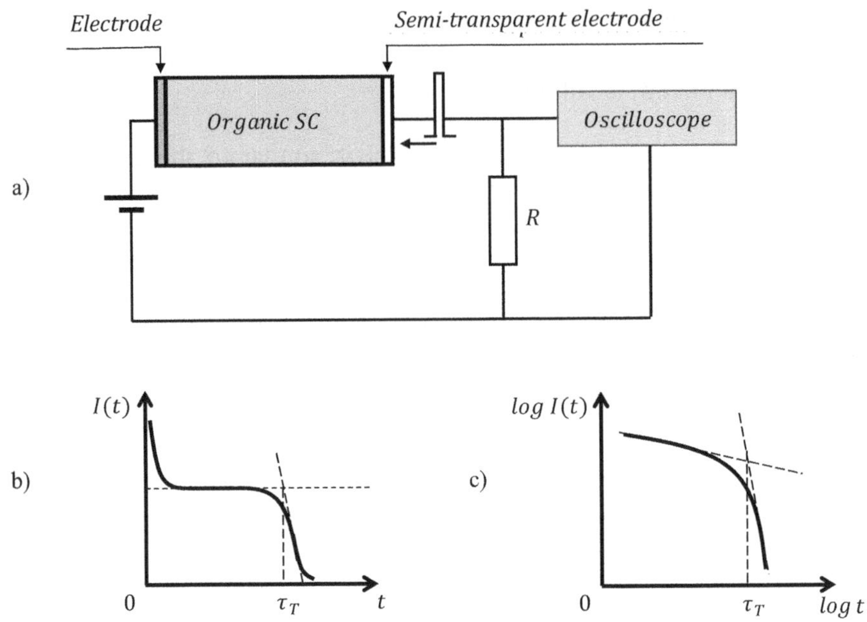

Figure 4.7. *a) Schematic setup for measuring time of flight; b) variation of non-dispersive photocurrent; c) variation of dispersive photocurrent on a double logarithmic scale*

For a *non-dispersive medium*, the charge carriers of a given type have the same drift velocity and therefore the transit time τ_T is determined by the inflection point of the curve $I(t)$, after which the current decreases.

For a dispersive medium, the movement of charges is governed by a time distribution and the curve $I(t)$ is in constant decline.

According to the Scher–Montroll theory (Scher and Montroll 1975), the transient current is given by:

$$t < \tau_T : I(t) \propto t^{-(1-\alpha)} \quad [4.20a]$$

$$t > \tau_T : I(t) \propto t^{-(1+\alpha)} \quad [4.20b]$$

In a double logarithmic scale, the transport time is determined by the inflection point of the curve $I(t)$.

If τ_T is known, we can determine the mobility of carriers with the help of the expression [4.19]:

$$\mu = \frac{d}{\tau_T\, E}$$

To realize the time-of-flight measurements, certain conditions must be met:

– the conductivity of the material must be low (specifically, the dielectric relaxation time of the material τ_σ must be greater than the transit time of the carriers $\tau_\sigma > \tau_T$);

– the duration of the light pulse must be shorter than the transit time;

– the energy of the light pulse must be greater than the gap of the SC in order to create electron–hole pairs;

– the absorption of the SC must be high, so that the photons can be completely absorbed by the sample. Its thickness should be sufficient (> 1 μm).

4.3.1.4.2. Charge extraction by linearly increasing voltage technique

The technique known as CELIV (charge extraction by linearly increasing voltage) is a variant of the time-of-flight technique which allows the mobility of the carriers to be determined, with the use of the measurement results being facilitated by the experimental conditions (Juska *et al.* 2000).

The principle for the measurement setup of this technique is given in Figure 4.8.

This setup provides for the photo-CELIV technique, which uses both a light excitation and a voltage ramp to create the charge carriers and extract them from the SC. The CELIV technique can be used with the voltage ramp alone, without the excitation of light (dark CELIV). In this case, the CELIV spectrum corresponds to the extraction of charges in equilibrium in the SC.

Depending on the type of carriers to be studied, the laser pulse can illuminate the metal electrode or the semi-transparent electrode. The voltage ramp will be applied in forward or reverse direction, depending on the carrier mobility to be measured.

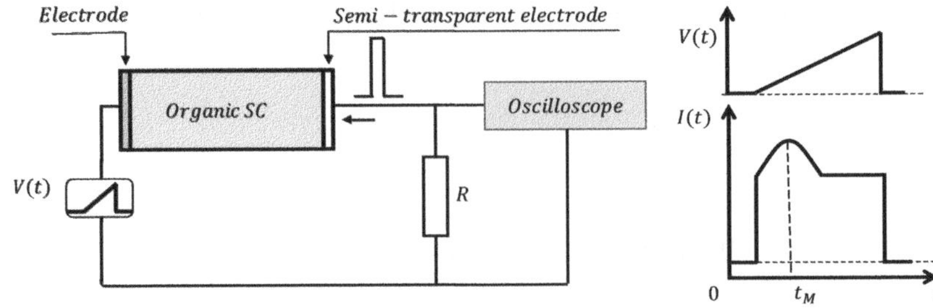

a) b)

Figure 4.8. *a) CELIV measurement setup; b) applied voltage ramp and CELIV spectrum recorded*

It is assumed that the charges of the SC are uniformly distributed throughout the volume of the sample and that we wish, for example, to measure the mobility of the electrons of an organic SC. The use of a hole-blocking electrode is necessary so that only the electrons from the SC are extracted. In this case, the semi-transparent electrode in the setup given in Figure 4.8 is hole-blocking. Conversely, an electron-blocking electrode should be used for measuring the hole mobility. The laser pulse creates electron–hole pairs in the SC that are assumed to be uniformly distributed within the volume of the sample. The application of the voltage ramp of the form $v(t) = A \times t$ allows electrons to be extracted and creates a current in the external circuit. The extraction of electrons creates a depleted area, which becomes wider when voltage is applied. When all the excess charges are extracted from the sample, the current decreases, approaching a limit that corresponds to the leakage current, called the *capacitive current.* On the CELIV spectrum, the current $I(t)$ is recorded as a function of time and a maximum current is observed at time t_M, called the extraction time, which represents the end of the extraction of excess electrons.

Assuming that there is no recombination and that the displacement of charges is the sole cause of the transient current, the current $I(t)$ is determined by Gauss's theorem, Kirchhoff's law and the continuity equation. From the expression of the current and the experimental value of t_M, we can deduce the expression for the mobility of the carriers, which is:

$$\mu = \frac{2\,d^2}{3\,A\,t_M^2} \qquad [4.21]$$

This expression is valid for SCs with low conductivity, that is, for cases where $\tau_\sigma > \tau_T$.

For materials with high conductivity, that is, $\tau_\sigma < \tau_T$, the mobility is given by:

$$\mu = \frac{d^2\,\tau_\sigma}{A\,t_M^3} \quad [4.22]$$

Finally, for an average conductivity $\tau_\sigma \sim \tau_T$, the mobility μ is calculated by the expression:

$$\mu = \mathrm{K}\,\frac{d^2}{A\,t_M^2\left[1+0{,}36\times\frac{\Delta j}{j(0)}\right]} \quad [4.23]$$

where K is a constant, Δj is the value of the maximum extraction current and $j(0)$ is the capacitive current.

The advantages of the CELIV technique with respect to the TOF technique are, first, the ability to directly measure the mobility of the charge carriers (holes and electrons) in SCs with a dispersive transport, and second, the convenience of using devices without transparent electrodes (dark CELIV) and without a minimum for the sample thickness to obtain the spectra.

4.3.2. *Space-charge limited conduction*

Space-charge limited conduction is often observed in organic electronic devices. It occurs in the region near an electrode for which the contact with the SC is ohmic. This electrode is theoretically capable of providing an unlimited amount of charge carriers to the SC, as long as the injection of these carriers is maintained by the application of an electric field.

If the applied voltage is low, the charges injected are less than the number of carriers in the SC at thermal equilibrium, and the current density follows Ohm's law, as given by:

$$J = N\,\mu\,E\,|e| = N\,\mu\,|e|\,\frac{V}{d} \quad [4.24]$$

where N is the charge carrier density.

In SCs with low electrical conductivity, when the number of carriers injected becomes greater than the intrinsic charges, the excess charges, which cannot be instantly discharged, accumulate near the contact electrode and form a *space charge region*. In this way, the injected charges control the space charge and the internal electric field of the SC. The movement of charges when a space charge is present is called the space-charge limited current (SCLC).

However, the SCLC that arises due to the ohmic contact between the SC and the electrode does not obey Ohm's law. In other words, the relationship between the current and the voltage is not linear. We can confirm this by considering a metal–polymer–metal diode whose polymer layer has a surface area S, thickness d and permittivity ε_r. The charge created in the diode when a voltage V is applied is:

$$Q = C\,V = \varepsilon_0\varepsilon_r\,\frac{S}{d}V$$

If this charge is uniformly distributed within the volume of the layer, the current density in the diode will be:

$$J = \frac{I}{S} = \frac{Q}{S\,\tau_T} = \varepsilon_0\varepsilon_r\,\frac{V}{d}\times\frac{\mu\,E}{d} = \varepsilon_0\varepsilon_r\,\frac{\mu\,E^2}{d}$$

where τ_T is the transit time and is defined by the expression [4.19].

In the case of an SCLC current, we know that the space charge created at the ohmic contact is not uniform. To obtain the expression for the current density in this case, the following equations are used:

– Poisson's equation:

$$-\frac{dE}{dx} = \frac{N|e|}{\varepsilon_0\varepsilon_r} \qquad [4.25]$$

– the continuity equation:

$$J = |e|N\,\mu\,E - |e|D\frac{dE}{dx} \qquad [4.26]$$

– the Einstein relation:

$$|e|D = \mu\, k\, T \qquad [4.27]$$

The calculation of the current density for an ideal SC (without defects) provides the following expression:

$$J = \frac{9}{8}\; \varepsilon_0 \varepsilon_r\, \mu\, \frac{V^2}{d^3} \qquad [4.28]$$

This expression is known as the *Mott–Gurney equation* or *Child's law*. It indicates a quadratic dependence of the current density as a function of the applied voltage. This density is also inversely proportional to the cube of the thickness of the sample and this dependence must be taken into account for the interpretation of the current–voltage characteristics in the SCLC process.

The transition voltage V_Ω between the ohmic regime and the SCLC regime is determined using expressions [4.24] and [4.28]. We obtain:

$$V_\Omega \;=\; \frac{8}{9}\, \frac{|e|\, N\, d^2}{\varepsilon_0\, \varepsilon_r} \qquad [4.29]$$

Expression [4.28] was obtained for an ideal SC without defects and it does not reflect the actual structure of the SCs, especially organic ones. These materials are considered to contain a large number of traps due to the nature of their molecular bonds and the inherent disorder of amorphous solids.

4.3.2.1. *The case of discrete traps centered on an energy level E_T*

Taking into consideration the effect of traps and assuming they are discrete, expression [4.28] for the density of the space-charge limited current becomes:

$$J = \frac{9}{8}\; \varepsilon_0 \varepsilon_r\, \mu\, \theta\, \frac{V^2}{d^3} \qquad [4.30]$$

In this expression, θ represents the portion of free charges in the SC. We can write:

$$\theta = \frac{N_{free}}{N_{free} + N_{trapped}} \qquad [4.31]$$

When traps are dominant, the conduction becomes *trap-limited*, or TL. In this case, the parameter θ is small and the majority of the charges injected will be trapped. Consider a set of electron traps with a total concentration of N_T occupying an energy level E_T in the SC. The concentration of trapped electrons is:

$$n_T = \frac{N_T}{1+exp\left(\frac{E_T-E_{Fn}}{k\,T}\right)} \quad [4.32]$$

The concentration of free electrons in the conduction band is:

$$n = N_C\, exp\left(\frac{E_{Fn}}{k\,T}\right) \quad [4.33]$$

From the expressions [4.32] and [4.33] and by establishing $\theta = \frac{N_C}{N_T} exp\left(\frac{E_T}{k\,T}\right)$, the following relationship is obtained:

$$n_T = \frac{N_T}{1+\frac{\theta\, N_T}{n}} \quad [4.34]$$

The proportion of free charges is:

$$\theta = \frac{N_C}{N_T} exp\left(\frac{E_T}{k\,T}\right) \quad [4.35]$$

For a low injection of charges, $n \ll \theta\, N_T$ and $n_T \sim n/\theta$, the expression of the current density [4.30] becomes:

$$J = \frac{9}{8}\,\frac{\varepsilon_0 \varepsilon_r\, \mu}{1+\frac{1}{\theta}}\,\frac{V^2}{d^3} \quad [4.36]$$

The current density is a quadratic function of the applied voltage and its value is lower than that of an SC without traps.

There are two types of traps, *shallow traps* and *deep traps*:

– For shallow traps, the trap level E_T is close to the quasi-Fermi level E_{Fn}: $E_T > E_{Fn}$. When these traps are filled, the charges injected after filling the traps are free to move. The current density evolves into a trap-free conduction, known as *trap-filled limited* (TFL) conduction. The threshold voltage V_{TFL} is defined as the voltage sufficient to fill the traps and represents the transition from the trap-filled regime to the trap-free conduction regime. Its expression is a function of the concentration of traps as follows:

$$V_{TFL} = \frac{|e|\, N_T\, d^2}{2\, \varepsilon_0 \varepsilon_r} \quad [4.37]$$

– For deep traps, given that $E_T < E_{Fn}$, the injected charges will all be trapped to fill the traps. The threshold voltage V_{TFL} represents the complete filling of the traps, and since all the traps are filled, V_{TFL} also represents the transition time from the ohmic regime to the trap-free conduction regime. In other words, the trap-filled regime and the transition to the trap-free regime are reduced to simply the transition from the ohmic regime to the trap-free conduction regime.

The different space-charge limited conduction regimes in the case of discrete traps of a concentration N_T and energy level E_T are summarized in Figure 4.9.

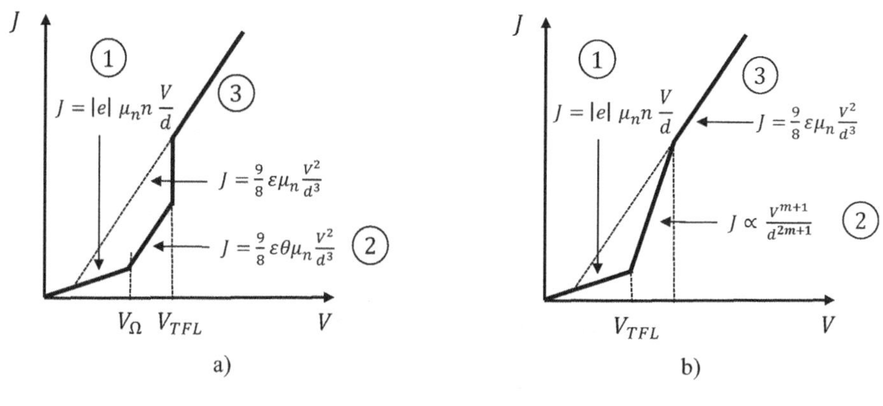

Figure 4.9. *The conduction regimes limited by space charges in an SC: a) containing shallow traps; b) containing deep traps. The different regimes are: (1) ohmic conduction; (2) trap-limited conduction; (3) trap-filled limited conduction*

4.3.2.2. *The case of an exponential distribution of traps*

In organic SCs, traps are rarely discrete and more generally distributed in energy. Therefore, Child's law can be modified. When an SC with an energy trap distribution is subjected to an applied electric field, the charge carriers injected will tend to fill the trap states starting with the lowest energy levels.

Consequently, as the electric field increases, the activation energy required to extract the trapped carriers decreases and the change in current density as a function of the applied field will be higher than that predicted by Child's law.

In an exponential distribution of traps, supposed to be electron traps, the distribution function can be written in the following form (Yang and Shen 1999):

$$N(E) = \frac{N_t}{E_t} exp\left(-\frac{E_C - E}{E_t}\right) \qquad [4.38]$$

where $N(E)$ is the concentration of trap states at the energy level E, N_t is the total concentration of traps, $E_t = k\,T_t$ is the energy characteristic of the distribution and E_C is the bottom level of the conduction band.

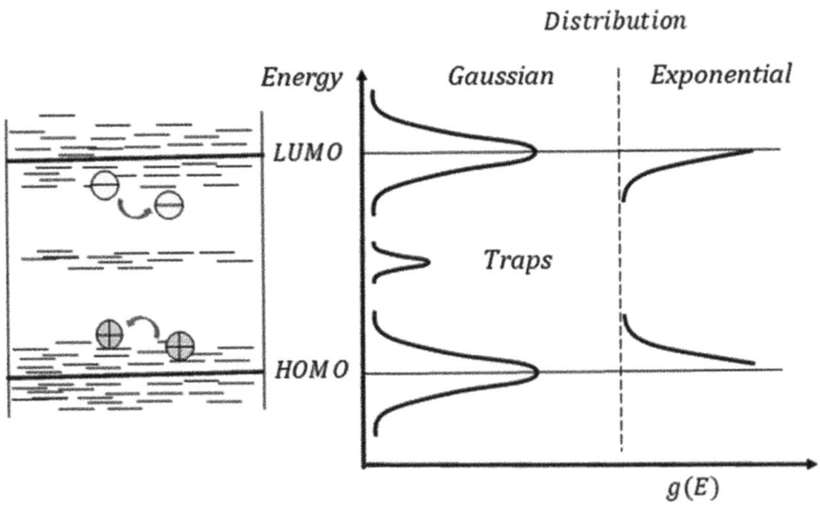

Figure 4.10. *Types of trap distributions: Gaussian and exponential*

The current density as a function of the applied voltage is obtained by the expression:

$$J(V) = |e|^{1-l} N_C\, \mu_n \left[\frac{\varepsilon_0\, \varepsilon_r\, l}{N_t(l+1)}\right]^l \left(\frac{2\,l+1}{l+1}\right)^{l+1} \frac{V^{l+1}}{d^{2l+1}} \qquad [4.39]$$

where N_C is the effective concentration of the conduction band and $l = \frac{T}{T_C}$.

The current density varies on the basis of the applied voltage as determined by the power law in the form $J(V) = KV^{l+1}$.

4.3.2.3. *The case of a Gaussian distribution of traps*

In this distribution, the energy states form a roughly Gaussian distribution and the charge carriers move by hopping between these states. The distribution contains both carrier types: free and trapped. The trapped carriers are located at the tail of the distribution where the density is low, while the free carriers are in the center where the density is high.

The distribution function can be written as:

$$N(E) = \frac{N_t}{\sqrt{2\,\pi}\sigma_t}\; exp\left[-\,\frac{(E-E_{tm})^2}{2\,\sigma_t^2}\right] \qquad [4.40]$$

where E_{tm} is the trap energy level of maximum concentration and σ_t is the standard deviation of the distribution.

For shallow traps, the Fermi level is located above E_{tm} and the current density as a function of the applied voltage is given by (Hwang and Kao 1976):

$$J = \frac{9}{8}\;\varepsilon_0\varepsilon_r\,\mu\,\theta\;\frac{V^2}{d^3} \qquad [4.41]$$

with:

$$\theta = \frac{N_C}{N_t}\exp\left[-\frac{E_t}{k\,T}+\frac{1}{2}\left(\frac{\sigma_t}{k\,T}\right)^2\right] \qquad [4.42]$$

For deep traps, the Fermi level is located below E_{tm} and the current density as a function of the applied voltage is similar to expression [4.39] for an exponential distribution of traps, but with an exponent value of $l = l'$, which is a function of the standard deviation σ_t as given by:

$$l' = \sqrt{1+2\,\pi\left(\frac{\sigma_t}{4\,k\,T}\right)^2} \qquad [4.43]$$

The approximations used to treat the Gaussian distributions are essentially the equations established for an exponential distribution or for a discrete number of traps. The current–voltage characteristics in a solid with traps distributed in a Gaussian manner are therefore similar to those of a discrete trap level at low voltages (i.e. a low density of occupied traps) and to that of an exponential trap distribution at high voltages (i.e. a high density of occupied traps).

Exponential and Gaussian distributions can be used experimentally to explain the current–voltage characteristics in devices based on various conjugated polymers without any notable contradictions (Nicolai *et al.* 2011).

4.3.3. *Defects and traps in organic semiconductors*

Defects in SCs play an important role in the optical and electronic processes of the material. Though it is possible to control the parameters of the material's preparation so as to improve its characteristics and performance, it is difficult to control the parameters of the defects of the same materials. On the one hand, defects are formed easily throughout the material, and on the other hand, it is not easy to accurately assess the parameters of the defects because the characterization methods are not easy to implement and the analysis of the measurements is complex.

In Chapter 1, we recalled the basic concepts of defects in inorganic SCs, covering their formation and their transitions, as well as some simple characterization techniques.

In this section, we will examine more specifically the defects in organic SCs that differ in nature from inorganic SCs and also the techniques of characterization that can be adapted for measuring the trap parameters.

4.3.3.1. *The origin of defects in organic semiconductors*

In inorganic SCs whose atoms are arranged in an ordered three-dimensional lattice, point defects are often created in the crystal lattice due to the lack of an atom at a lattice site (*vacancy*) or the incorporation of a foreign atom so as to occupy a vacant lattice site (*substitutional impurity*). If it is located in the empty space of the lattice, this is called *interstitial impurity*. In organic SCs, the molecules are distributed randomly, without any special orientation and in a disorderly state within the volume of the material. This peculiar characteristic gives rise to a less compact structure with the empty spaces between the molecules, and as a result, there are more opportunities for defects to form in these materials.

From a chemical point of view, inorganic SCs are formed from the elements of column IV of the periodic table. The bonds between atoms are covalent bonds that ensure the solidity and stability of the chemical structure. As a result, SCs are not chemically affected by the surrounding environment in which they are placed. For organic SCs, the molecules are bound together by van der Waals bonds that are weak and can be easily broken. Thus, the SCs are strongly affected by the physical conditions (such as temperature) and chemical conditions (such as the presence of oxygen) of the environment, because they allow reactions with open bonds and thus modify the structure of the material.

The two previous causes can also be combined to effectively give rise to the formation of defects in the material. Due to the weak mechanical properties and lower compactness of organic materials, the diffusion of impurity atoms and molecules is facilitated in the layers, creating defects throughout the sample, while this diffusion is limited in inorganic SCs due to their compact and organized structure.

The defects noted above consist of the imperfections in the structure of the material itself, which can be referred to as intrinsic defects. However, in an electronic device, the organic layer is brought into contact with other layers to ensure the charge transport and to fulfill its function. Given the nature of the bonds of the material, the interface between the organic layer and a contact layer can be expected to be the source of other defects in addition to intrinsic defects. As we will see in the next chapter, the morphology of the films has a great impact on the structure of the interface, and therefore on the charge injection processes and ultimately on the transport and performance of the devices in which they are used. An interface containing a high concentration of defects significantly influences the physical processes in the material, and therefore must be avoided in principle.

4.3.3.2. *Study methodology of defects in organic SCs*

Now that we have examined the causes of how defects are formed, we can establish the methods for studying defects, considering the possible ways in which they are formed in the material and in devices.

First of all, the inherent defects of the material can be studied using analytical techniques that will be recalled or described in the next section. To better identify the causes of how they are formed, the material can be subjected to different experimental conditions by changing the parameters believed to be factors in the development of defects. To study the influence of light and moisture on the environment, organic samples are frequently tested by exposing them to different storage conditions (in air, in a vacuum, in bright light, in the dark and through a combination of these conditions) and by carrying out characterization methods to identify the physical properties impacted by the formation of defects. Based on the analysis of the defect parameters, we can reach conclusions on the processes of formation and the possible causes of these formations. As an example, one study conducted on polyfluorene (PF) and the composite PF/ZnO (Zou *et al.* 2011) based on this methodology and using optical spectroscopic techniques allowed conclusions to be reached on the role of light and oxygen in the creation of defects through the oxidation of the polymer. It also showed that the use of ZnO oxide nanoparticles helped to mitigate the effect of photo-oxidation and maintain the stability of the composite.

The study of intrinsic defects provides information on the organic material considered in isolation. In no case does it allow an accurate determination of the parameters of the defects in the electronic device, for the reasons previously given. In other words, there would be greater risk in extrapolating the results obtained from the measurements made on the materials alone onto those obtained in a device using the same material. With the exception of a few special cases, it is generally accepted that the contact between one organic material and another organic material or with a metal will alter the structure of the surface of the original material, and as a result, will create additional defects in the device. Defects in the interface can contribute to intrinsic defects, change them or even become dominant. Thus, if the measurements of the parameters of the defects are carried out on a single device, based on the trap parameters obtained, we cannot expect to accurately identify the processes that form them, nor the role of each element involved in these processes.

In order to distinguish the role of each element involved in forming the defects of organic materials, we can proceed to study the defects in the device by removing or adding the involved element. This method consists of determining the defect parameters in devices that are identical except for the fact that they have or have not included the intervening element in question. If the parameters obtained are identical with or without this element, it does not play any part in the process of creating the defects. This method allows us to determine the role played by each component involved in the overall process of creating defects in a device.

For example, for the investigation of the defects in polyfluorene-based light-emitting diodes with different diode structures (Renaud and Nguyen 2010), different operating conditions (Renaud and Nguyen 2009) and different compositions of layers (Renaud and Nguyen 2008; Nguyen and Renaud 2009) allowed a better understanding of how defects are formed and enabled the determination of the nature of such defects in these devices.

4.3.3.3. *Measurements of trap parameters*

The techniques for the characterization of traps were presented in Chapter 1. In this part, we will study the measurements of trap parameters in organic SCs, along with the most frequently used techniques as well as those best suited to the nature of these materials.

A trap is an *electrically active defect*, that is, as a result of its charge, it is capable of interacting with a mobile charge carrier so as to hold it for a certain time, thus preventing it from moving freely and participating in conduction. The carriers caught by a trap can be released from the trap if they receive a sufficient amount of energy (thermal, electrical or light energy).

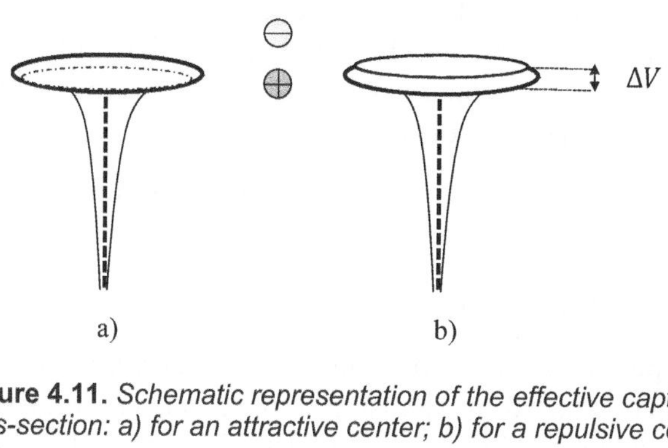

Figure 4.11. *Schematic representation of the effective capture cross-section: a) for an attractive center; b) for a repulsive center*

First, we will recall the trap parameters used to characterize the defects of the SC. Each discrete level of the trap can be characterized by the following three elements:

– the *energy level* E_T in the forbidden band of the SC. It is located with respect to the allowed band edge for either E_C for the CB or E_V for the VB. The difference in energy between E_T and E_C (or E_V) constitutes the *activation energy* for the process of releasing the captured carrier. This energy is often used to identify and distinguish *shallow traps* from *deep traps*. However, the values chosen for this energy are often arbitrary, and it is preferable to situate these levels in relation to a known energy level, such as the Fermi level E_F. For intrinsic SCs or doped SCs, the quasi-Fermi level E_{Fn} (N-type SC) or E_{Fp} (P-type SC) is used. The traps are shallow if the energy level of the trap is above the quasi-Fermi level E_{Fn} for an electron trap and below the quasi-Fermi level E_{Fp} for a hole trap. Conversely, deep traps are located on the opposite side of that of the shallow traps with respect to the same quasi-Fermi levels;

– the *capture cross-section* σ (in cm^2). This parameter has a surface dimension and represents a *virtual* surface surrounding the trap. This surface delimits the space within which a charge carrier will be captured by the trap. The capture cross-section expresses the probability that a carrier will be captured, which depends on the electrostatic interaction between the charges of the trap and the carrier. For an *attractive center*, this probability will be high (high value of σ) and the charge carrier is easily captured. For a *repulsive center*, σ is small, and the charge carrier has little chance of being trapped, since it must overcome a potential barrier ΔV (of repulsion) in order to interact with the charge of the trap;

– the *density* N_T, which represents the number of traps per unit volume of material.

4.3.3.3.1. Thermally stimulated current technique

The principle for measurement using the *thermally stimulated current* (TSC) technique is based on releasing trapped carriers through a controlled thermal input and measuring the current associated with these carriers during the de-trapping process. A schematic diagram of the technique is given in Figure 4.12.

The basic measurement consists of the following steps:

– step 1: a voltage V is applied at a high temperature T_H to the sample, allowing charges to be injected into the active layer and captured by the traps;

– step 2: while maintaining the voltage V at the terminals of the sample, the temperature of the sample is lowered to a low temperature T_L to freeze the trapped carriers. The sample is then short circuited in order to discharge all residual free charges, except those captured by the traps. The system returns to thermal equilibrium after a certain time;

– step 3: the sample is heated at a constant rate β and the current generated from the release of trapped carriers is recorded as a function of time. The spectrum obtained is the TSC current, and the analysis of this current allows the parameters of the sample's traps to be obtained.

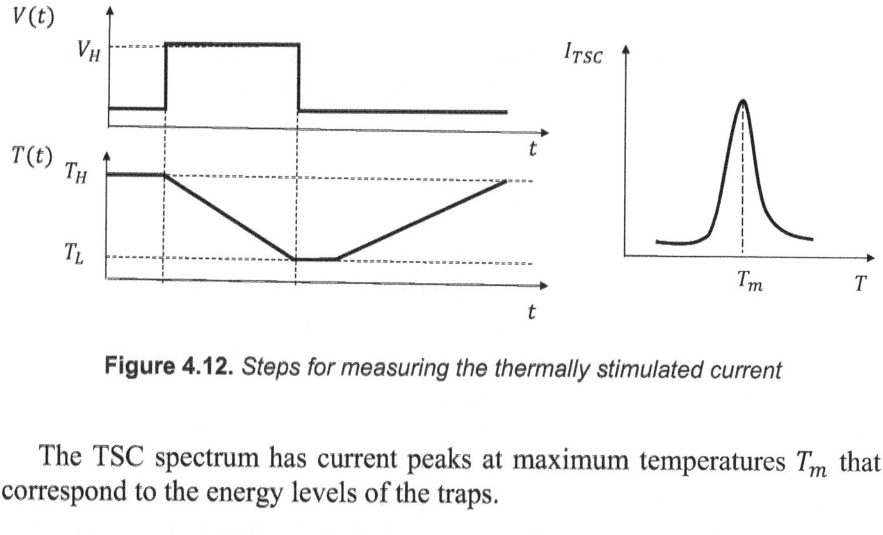

Figure 4.12. *Steps for measuring the thermally stimulated current*

The TSC spectrum has current peaks at maximum temperatures T_m that correspond to the energy levels of the traps.

Assuming that the emission of the trapped carriers obeys Boltzmann statistics, the TSC current is written as:

$$I_{TSC} = -C\frac{dN_T}{dT} = C\nu N_T \exp\left(-\frac{E_T}{k\,T}\right) \qquad [4.44]$$

In this expression, C is a coefficient of proportionality and ν is the attempt-to-escape frequency of trapped carriers.

For a constant heating rate β, the expression of the TSC current is given by the Cowell–Woods formula (Cowell and Woods 1967):

$$I_{TSC}(\mathrm{T}) = N_{t0}\ \nu \exp\left(-\frac{E_T}{k\,T}\right) \times \exp\left(-\frac{\nu}{\beta}\right) \int_{T_0}^{T} \exp\left(-\frac{E_T}{k\,T'}\right) dT' \qquad [4.45]$$

where N_{t0} and T_0 are, respectively, the initial concentration and temperature of the trapped charges.

For temperatures below T_m, the current I_{TSC} is proportional to $exp\left(-\frac{E_T}{k\,T}\right)$. The activation energy is determined by the slope of the Arrhenius curve $ln(I_{TSC}) = f\left(\frac{1}{k\,T}\right)$. This method is called the *initial rise method.* It is frequently used for determining the trap depth in SC-based devices.

The concentration of the traps is evaluated by integrating the current curve I_{TSC} to obtain the amount of relaxed charges as given by:

$$N_T = \frac{\int_{T_L}^{T_H} I_{TSC}(T)dT}{|e|\,S\,d} = \frac{\Delta Q}{|e|\,S\,d} \qquad [4.46]$$

The basic TSC technique is used to determine the parameters of the traps associated with a discrete level or the principal level of a distribution of the traps. We can also use a more refined method to study a trap distribution. This method, known as *fractional TSC*, consists of partial emptying of the traps by temperature cycles. Each cycle consists of a rise in temperature and then a cooling to the initial temperature T_L. The relaxed current is recorded as a function of the temperature. By performing several cycles at regular temperature intervals, the activation energies and trap concentrations are determined by recomposing each fractional spectrum using the Cowell–Woods formula. Then, we deduce the trap energy distribution $N_T = f(E_T)$.

The TSC technique is frequently used to study defects in organic SCs and makes it possible to establish the trap profiles in the gap of these materials. For example, light-emitting diodes or organic solar cells (Schafferhans *et al.* 2010) have been studied using this technique.

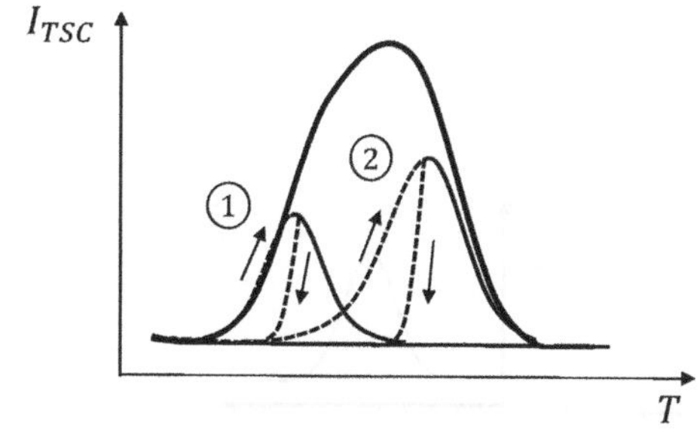

Figure 4.13. *The principle of fractional TSC current measurement: temperature cycles illustrating the heating and cooling steps (dotted line) and TSC spectra calculated with the Cowell–Woods formula (solid line)*

4.3.3.3.2. Deep-level transient spectroscopy

The conventional version of this technique and its application to inorganic SCs were described in Chapter 1. In this version, the change in capacitance ΔC as a function of the time of the sample is measured after the application of a voltage pulse.

For organic SCs in which the mobility of the carriers is generally low, the method does not allow exploitable spectra, as the variations in capacitance ΔC remain very small. To overcome this issue, a variant of the technique known as *charge-based deep-level transient spectroscopy* (Q-DLTS) is used. Instead of the capacitance, measurements are taken of the changes in the charge of the sample, using the same protocol as the conventional technique.

The principle of Q-DLTS is shown in Figure 4.13.

For measurements, a pulse of voltage V and width t_C (called the *charging time*) is applied to the sample to fill the traps. At the end of the pulse, the voltage is set to zero, and the captured carriers are released through thermal agitation. A measurement is taken of the amount of charge $\Delta Q(t)$ that is relaxed between two times t_1 and t_2, and the curve $\Delta Q(t)$ is plotted as a function of the variable called the *time window* defined by the expression:

$$\tau = \frac{t_2 - t_1}{ln(t_2/t_1)} \qquad [4.47]$$

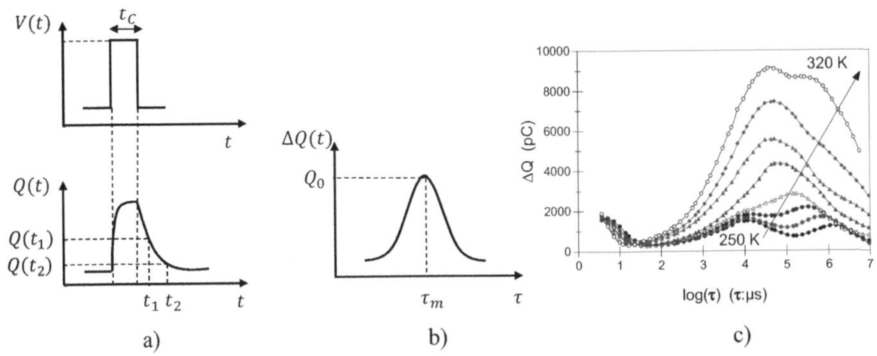

Figure 4.14. *The principle of charge-based deep-level transient spectroscopy: a) the pulse used and the charge of the sample; b) the Q-DLTS spectrum at a constant temperature; c) experimental Q-DLTS spectra at different temperatures of a perovskite-based solar cell. For a color version of this figure, see www.iste.co.uk/nguyen/electronics1.zip*

This curve is the Q-DLTS spectrum and contains peaks of charge values Q_0 that correspond to the energy levels of the traps of the sample. For these values, the corresponding time window is equal to the relaxation time of the captured carriers or the inverse of the emission rate e_{pn} of holes or electrons by the traps:

$$\tau_m = \frac{1}{e_{pn}}$$

The amount of charges relaxed between the two time points t_1 and t_2 is equal to:

$$\Delta Q = Q(t_1) - Q(t_2) = Q_0\left[\exp\left(-e_{pn}t_1\right) - exp\left(-e_{pn}t_2\right)\right] \qquad [4.48]$$

In practice, the times t_1 and t_2 are chosen to obtain a constant ratio between these two times: $\alpha = t_2/t_1 = 2$. The change in charge ΔQ as a function of the time window is written under these conditions:

$$\Delta Q = Q_0\left[\exp\left(-\ln(2)\frac{\tau}{\tau_m}\right) - \exp\left(-2ln(2)\frac{\tau}{\tau_m}\right)\right] \qquad [4.49]$$

By varying the temperature of the sample, it is possible to obtain for each charge peak the activation energy E_T of the trap center and the effective capture cross-section σ by plotting the Arrhenius curve $\left[-\ln\left(\tau_m T^2\right)\right]$ as a function of $\left(\frac{1}{k\,T}\right)$. The concentration of traps is also obtained by the expression:

$$N_T = \frac{4\,Q_0}{|e|\,S\,d} \qquad [4.50]$$

The Q-DLTS technique has been successfully used to determine the trap parameters of light-emitting diodes (Nguyen *et al.* 2004) and organic solar cells (Nguyen *et al.* 2012).

4.3.3.3.3. Impedance spectroscopy

Impedance spectroscopy is frequently used to study defects in electronic devices. It consists of measuring the frequency responses of the device when an alternating current AC of frequency v is applied to it (the corresponding angular frequency is $\omega = 2\,\pi v$). A direct voltage V_{DC} can also be superimposed onto the AC voltage to create a charge distribution that is utilized to study defects.

The detection of defects is based on the measurement of the capacitance C_S of the Schottky contact at the interface between the SC and the metal electrode:

$$C_S = \varepsilon_0 \varepsilon_r \frac{S}{W} \qquad [4.51]$$

In this expression, W is the thickness of the space charge region of the contact. At low frequencies, the charge carriers in this region can be captured by the defects or released from these defects depending on the applied signal alternation. These charge movements contribute to the total capacitance of the sample. At high frequencies, the emission rate of the carriers is too low with respect to that of the alternating signal and can no longer follow the changes in the applied voltage. As a result, the charges captured or released by the defects no longer contribute to the measured capacitance $C(\omega)$ that tends towards the geometric value of the sample:

$$C_0 = \varepsilon_0 \varepsilon_r \frac{S}{d} \qquad [4.52]$$

Figure 4.14 shows the typical variation in capacitance $C(\omega)$ as a function of frequency. At low frequencies, the capacitance reaches a plateau with a constant permittivity ε_s, known as static permittivity. At high frequencies, a plateau with constant permittivity $\varepsilon_\infty = \varepsilon_r$ can also be observed.

The transition frequency ν_0 between two active and non-active regimes of traps of energy E_T corresponds to the emission rate e_{pn} of the trapped carriers. This frequency is determined from the characteristic $C(\omega)$ by calculating the derivative of the capacitance $\omega\frac{dC}{d\omega}$. The angular frequency ω_0 corresponds to the maximum of the derivative and the concentration of defects N_T corresponds to the maximum variation $\Delta C(\omega_0)$. If the material has several levels of traps, the derivative $\omega\frac{dC}{d\omega}$ curve will present several peaks, whose parameters will be determined by the same analysis.

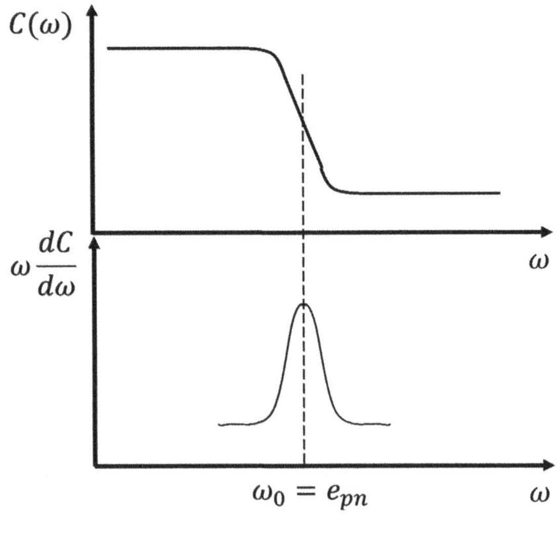

Figure 4.15. *Variation of capacitance $C(\omega)$ and its derivative $\omega\frac{dC}{d\omega}$ as a function of frequency*

Usually, the frequency response of the sample can be represented by an equivalent circuit that includes resistive and capacitive elements. For a sample without defects, an equivalent circuit has a resistance of R_S connected in series with a parallel circuit (R_p,C_p).

When discrete traps with energy of E_T are present, their effect on the frequency response can be represented by a series circuit (R_s,C_s). If the traps are distributed with a function of $f_t(E)$, the capacitance arising from the charges in the traps can be calculated using the following expression (Burtone *et al.* 2012):

$$C(\omega) = \frac{|e|^2}{k\,T} \int_{E_V}^{E_C} \frac{g\,(E)\, f_t(E)[1-f_t(E)\,]}{1+j\omega\,/\omega_t}\, dE \qquad [4.53]$$

where ω_t is the trap frequency, $f_t(E)$ is the distribution function and $g(t)$ is the density of states.

Software is available for the adjustment of experimental curves by simple equivalent circuits to interpret the effects of defects on the response to the alternative signal of the device under study.

The concentration of defects can be determined from the measurements of the capacitance-superimposed direct voltage characteristics $C(V_{DC})$. Effectively, the expression for Schottky capacitance without traps is written as:

$$C(\omega) = S\sqrt{\frac{|e|\,\varepsilon_0\,\varepsilon_r}{2} \times \frac{|N_A - N_D|}{V_{bi} - V_{DC} - k\,T/|e|}} \qquad [4.54]$$

The slope of the curve $\frac{1}{C^2}$ as a function of the applied direct voltage V_{DC} is proportional to the concentration $(N_A - N_D)$, and the intercept with the x-axis is equal to $V_{bi} - kT/|e|$.

When traps are present, the concentration of the carriers is changed by adding the charges arising from the traps. Let N_T be the concentration of traps assumed to be hole traps. The slope of the curve $\frac{1}{C^2}(V_{DC})$ will be proportional to $(N_A + N_T - N_D)$ (see Figure 4.15). The curve has two portions of different slopes that correspond to the conduction regimes with and without traps.

Impedance spectroscopy is widely used to study defects in light-emitting diodes (Campbell *et al.* 2000) and organic solar cells (Bisquert *et al.* 2008).

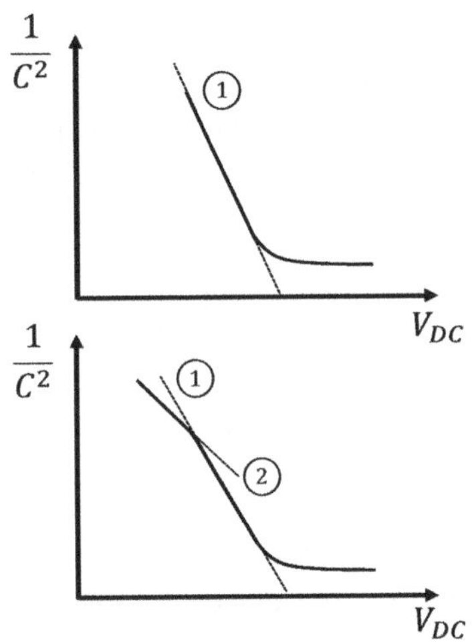

Figure 4.16. *Variation of $\frac{1}{C^2}$ as a function of the direct voltage V_{DC}. 1) trap-free regime; 2) trap regime*

4.3.3.3.4. Numerical or computer simulation of charge transport

Several variants of the methods presented above exist which are based on the same physical principle but which use different techniques for detection and/or measurement. For example, for deep-level transient spectroscopy (DLTS), such methods include the optical DLTS technique, deep-level optical spectroscopy (DLOS), which uses a light source to vary the photo-capacitance of the sample (Nakano *et al.* 2006). Similarly, when transient current is measured instead of capacitance, the DLTS by current (I-DLTS) technique is used (Neugebauer *et al.* 2012).

In addition to these techniques, there is a complementary method that does not rely on a particular trap capture or release process and does not use any specific devices to determine the trap parameters. This method is based on the numerical or computer simulation of the transport. It consists of establishing the transport equations adapted to the conditions of use of the devices and solving them, taking into account the boundary conditions, so as to theoretically deduce the variations from the simple characteristics which are easily accessible from the experimental measurements. The basic transport equations are the equations of the drift current, the diffusion current, the continuity equation and Poisson's equation. To study the traps, the characteristic of the device without a trap is first established, then the traps are introduced with a distribution type and the parameters of the distribution are adjusted so as to obtain the correct adjustment between the simulated characteristics and the experimental characteristic (Knapp *et al.* 2010).

The simulation technique of the charge transport is also applied in order to study physical processes (Blades and Walker 2000) other than trapping, which we will not expand on here.

5

Interface Processes

5.1. Introduction

In the previous chapters, we discussed the role of interfaces in the optical and transport processes that occur in organic electronic devices. An organic electronic device is obtained through successive depositions of thin metallic and organic layers on conductive or insulating substrates. Such designs involve several interfaces of different natures and structures, and the physical processes essential for device operation are influenced by the presence of these interfaces. The formation of such interfaces is inevitable in all inorganic and organic SC components. Understanding the interface processes is necessary to improve the conditions under which they are formed, and by consequence, the performance of devices. In this chapter, we will examine the surface characteristics of the materials used, the formation of interfaces between different materials, including metals in electrodes, and the effects these have on optical and electronic processes.

5.2. Formation of organic semiconductor/metal interfaces

In the study of inorganic SC-based junctions, the basic assumption for how a contact is formed between two SCs is that their Fermi energy levels are in alignment. This assumption, while well-founded, remains controversial due to the fact that it has been subject to limited experimental verification on different SC materials and metals. New concepts and assumptions have been developed to explain the formation and properties of the SC/metal interface, which can be extended to other interfaces formed between organic SCs and other materials.

5.2.1. *Vacuum-level alignment model: Mott–Schottky theory*

In this model, it is assumed that the vacuum levels of the materials in contact are in alignment. The vacuum level refers to the energy level of an electron at rest in the vacuum, which is far enough from the surface of the material that it does not influence that electron. This vacuum level is assumed to be constant for all materials and to extend to infinity (Ishii *et al.* 1999).

Given these considerations, according to the Mott–Schottky theory (Mönch 1990), when a metal is deposited on the surface of an inorganic SC, the Fermi levels of the materials align at thermodynamic equilibrium, as in the case of the formation of a P–N semiconductor junction. This alignment leads to the formation of a potential barrier at the SC/metal interface, generally known as the Schottky barrier.

For an N-type SC with electron affinity $|e|\chi_S$, the electrons from the CB migrate towards the metal with a work function $|e|\Phi_M$, leaving a zone of positive fixed charges of a concentration $N_D - N_A$. On the side of the metal, excess electrons accumulate close to the area where physical contact occurs, and with the positive charges of the SC, they create a space charge region (SCR) associated with a potential barrier $|e|\Phi_{bn}$:

$$|e|\,\Phi_{bn} = |e|\,(\Phi_m - \chi_S) \qquad [5.1]$$

Thus, the potential barrier depends on the work function of the metal, and the choice of the electrode metal will determine the charge injection for this model. The curve representing the change of $|e|\Phi_{bn}$ as a function of $|e|\Phi_m$ is a straight line with a slope $S = \frac{d(|e|\Phi_{bn})}{d(|e|\Phi_m)} = 1$.

As we discussed in Chapter 4, changes in the current voltage characteristics are only experimentally observed in very few systems, following expression [5.1].

5.2.2. *Interface dipole model: Bardeen's theory*

To explain why the $J(V)$ characteristics are relatively independent of the nature of the electrode metal, Bardeen's theory (Bardeen 1947) suggests that the concentration of charges in the SC/metal interface is altered by *interface states* that are created by the defects in this region. These defects have two origins: first, the disruption of the periodicity on the surface of the crystalline SC that creates dangling bonds, and second, the rough morphology of the sample.

To characterize these interface states, the *charge neutrality level* $|e|\Phi_0$ is introduced and defined as the highest energy level of the occupied interface states. This level is defined with regard to the band edge E_V of the VB. Depending on their positioning relative to the Fermi level E_F of the SC, the interface states can be occupied or empty. They are *donors* if they are positively charged when empty and neutral when occupied. Similarly, these states are *acceptors* if they are neutral when empty and negatively charged when occupied. On both sides of the level $|e|\Phi_0$, the acceptors and donors balance each other, which means that the total charge of the interface region remains neutral.

If the density of states of the interface around the level $|e|\Phi_0$ is high, the Fermi level of the SC approaches the $|e|\Phi_0$ level. The charge exchange between the metal and the interface states allows an *interface dipole* to be formed, with a shift ΔE of the vacuum level, and the Fermi level of the metal and the level of the surface states to align (see Figure 5.1).

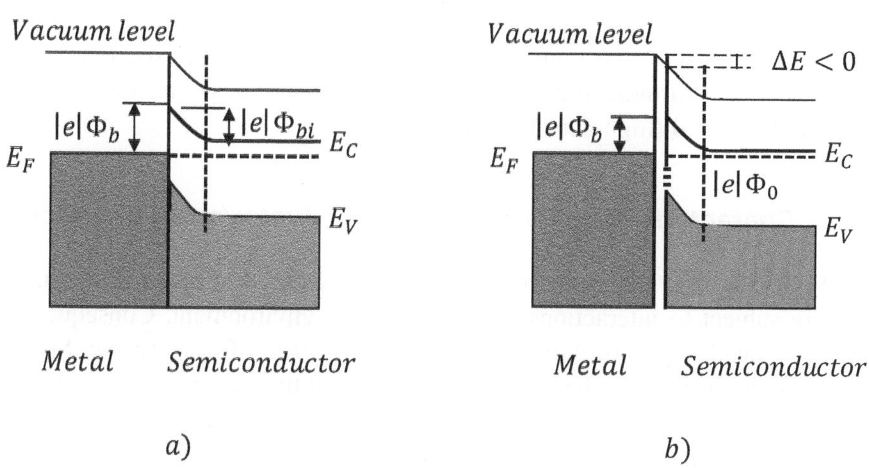

Figure 5.1. *Schematic representation of an N-type semiconductor/metal contact according to: a) the Mott–Schottky model; b) the Bardeen model. For a color version of this figure, see www.iste.co.uk/nguyen/electronics1.zip*

The potential barrier for the flow of electrons from the CB of the SC to the Fermi level of the metal becomes independent of the work function of the metal, and depends only on the interface states. Thus, this behavior is known as *Fermi-level pinning*, meaning that the Fermi level is pinned to the level $|e|\Phi_0$ (Braun *et al.* 2009).

When Fermi-level pinning occurs, the Schottky barrier is given by:

$$|e|\Phi_{bn} = E_G - |e|\Phi_0 \quad [5.2]$$

This expression clearly indicates that the potential barrier no longer depends on the work function of the metal because the Fermi levels of the metal and the SC are in alignment with the neutral energy level $|e|\Phi_0$.

To take into account the surface states and the nature of the metal, the expression of the potential barrier is given in the form of (Gao 2010):

$$|e|\Phi_{bn} = |e|(\Phi_m - \chi_S) - \Delta E \quad [5.3]$$

By introducing an interface parameter (or slope parameter) S, defined as the derivative of Φ_{bn} with respect to Φ_m:

$$|e|\Phi_{bn} = S\,|e|\,(\Phi_m - \chi_S) + (1 - S)(E_G - |e|\Phi_0) \quad [5.4]$$

On the basis of the density of states of the interface, we obtain the Mott–Schottky formula for $S = 1$ (low density of states) and the Bardeen formula for $S = 0$ (high density of states).

5.2.3. *Characteristics of organic semiconductor/metal interfaces*

With respect to inorganic materials, organic SCs are structurally unstable and are subject to interactions with the external environment. Consequently, when a metal layer is deposited onto the surface of an organic SC by evaporation, the reaction between the molecules of the SC and the high-energy metal atoms leads to chemical reactions at the interface between the two materials. These interactions are often observed at the interfaces between an inorganic SC and a metal electrode, which result in the formation of SC compounds. Many studies of organic SC/metal interfaces have been carried out experimentally and have shown the chemical reactions between the materials (Ip *et al.* 2003). In these cases, an interface layer is formed with a different nature and composition from that of the SC and the metal. The change in the electronic structure of the materials is observed by the modifications of the chemical environment of the atoms, showing that reactions occur as a result of the deposition of the metal atoms on the surface of the organic SC. These interactions have been the subject of theoretical investigations predicting changes in the electronic structure of the contact region (Brazowskii and Kirova 1993). The approach of this study is based on the concept of the minimum free energy of the SC/metal system at thermodynamic equilibrium: when the contact is made and the thermal equilibrium of the system is established, the charge transfer between the metal and the SC takes place, and polarons and bipolarons are formed in the interface region. The charge density and potential distribution in this region can be established by minimizing the free energy of the system. Thus, the electronic structure of the contact region can be determined. From a structural point of view, the formation of the interface has also been examined by studying the possible bonds between the metal atoms and the molecules of the SC and by determining the most favorable energy conditions using calculation methods from quantum mechanics. The Schrödinger equation cannot be thoroughly solved for systems consisting of interface atoms, and further approximations need to be used. The minimum energy of the system corresponds to the optimal geometry of the atoms, which is the location of the preferred placement of atom sites and the conformation of the organic material. This optimal geometry makes it possible to specify and describe the reaction between metal atoms and organic molecules. Several studies have been carried out on the polymer/metal (Brédas *et al.* 1982) and small molecule/metal (Curioni and Andreoni 2000) interfaces, the results of which have been confirmed by measurements using UV photoelectron spectroscopy (UPS).

In the analysis of organic SC/metal interfaces, it has also been observed that a diffusion of metal atoms from the electrode to the SC often occurs after they are formed (Nguyen and Ip 2002). The interface acts as an intermediate layer where the diffused atoms are formed and propagate. This layer is especially important for the operation and stability of devices.

The formation of the organic SC/metal interface also depends on the conditions under which the organic material has been prepared. While the small molecules and oligomers can be obtained through thermal evaporation, the polymer films are often prepared through spin-coating from solutions. The conditions of the deposition must be strictly controlled to prevent the deposited film from oxidizing, as well as to prevent impurities from forming on the film surface. The same process applies to the deposition of the metal film, the surface of which can be oxidized or contaminated, altering the structure of the interface.

5.2.4. *Fermi-level pinning*

According to the Mott–Schottky theory, the potential barriers at the SC/metal interface can be determined by equation [5.1], with the Fermi levels of the materials in alignment. The vacuum level can also be examined from the position of the level E_F of the contact that is formed. Since the Fermi level is constant across the SC/metal system, the vacuum level (VL) can be determined if the work function $|e|\Phi_m$ of the metal is known. We can then express the energy levels at the band edges with respect to the Fermi level E_F as:

$$E_V - E_F = |e|\,\Phi_m - E_i \qquad [5.5a]$$

$$E_C - E_F = |e|\,\Phi_m - |e|\chi_s \qquad [5.5b]$$

where E_i (IE or IP) is the ionization energy of the SC.

For a metal with a work function $|e|\Phi'_m > |e|\Phi_m$, the vacuum level will be higher than that of a metal with a work function $|e|\Phi_m$. As a result, the levels E_V and E_C are shifted towards the vacuum level, as shown in Figure 5.2.

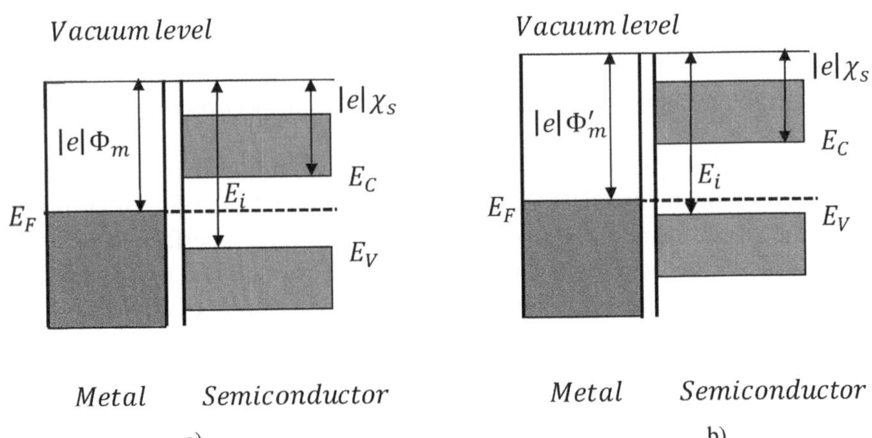

Figure 5.2. *Diagram of the energy bands of the metal/SC contact with alignment of vacuum levels: a) metal with the work function $|e|\,\Phi_m$; b) metal with the work function $|e|\Phi'_m$. For a color version of this figure, see www.iste.co.uk/nguyen/electronics1.zip*

If the work function of the metal becomes very high, the level E_V would be located above the Fermi level, thus allowing a rearrangement of the charge distribution in the contact region to occur, since the occupation of the energy states of the CB and the VB depends on the position of E_F. This reorganization of the states gives rise to the pinning of the metal work function to the energy level of the CB or VB.

The process of Fermi-level pinning takes place with the fixing of the energy levels when the Fermi level E_F crosses the distribution of the density of states (DOS) at the interface of two materials. This process occurs at the contact between two materials, regardless of the type of materials (organic or inorganic).

The process can be explained in the following way: before contact, the vacuum levels of the materials are aligned, and the energy states of each material are occupied according to the scheme that is already known. When the contact is made, the alignment of the energy levels is no longer at the vacuum level, but at the surface potential level. In other words, the occupied states of the VB of an SC that are situated above E_F before contact will empty after contact, as described by the Fermi–Dirac statistic. Conversely, the unoccupied states of the CB that are located below E_F before contact will be filled, as given by the same statistic. According to this scheme, a charge transfer occurs through the interface formed by the two materials: the electrons pass into the CB, and the holes into the VB. When the system is in equilibrium, the transfer of charges ends, and the energy levels of these states are set at a level called the *pinning level*, which is close to the Fermi level of the system. As a result, dipoles are created at the interface, as shown in Figure 5.3.

If the metal has a low work function, the alignment of the vacuum levels positions the level E_C above the Fermi level E_F. The transfer of charges creates a dipolar interface layer with a shift ΔE between the energy vacuum levels (see Figure 5.3a). For a metal with a high work function, the energy levels of the SC do not change, but the energy shift of the vacuum levels is greater (see Figure 5.3b). The potential barrier with the interface dipole layer for the electrons is given by expression [5.3] and depends on ΔE, which characterizes the dipole states of the surface.

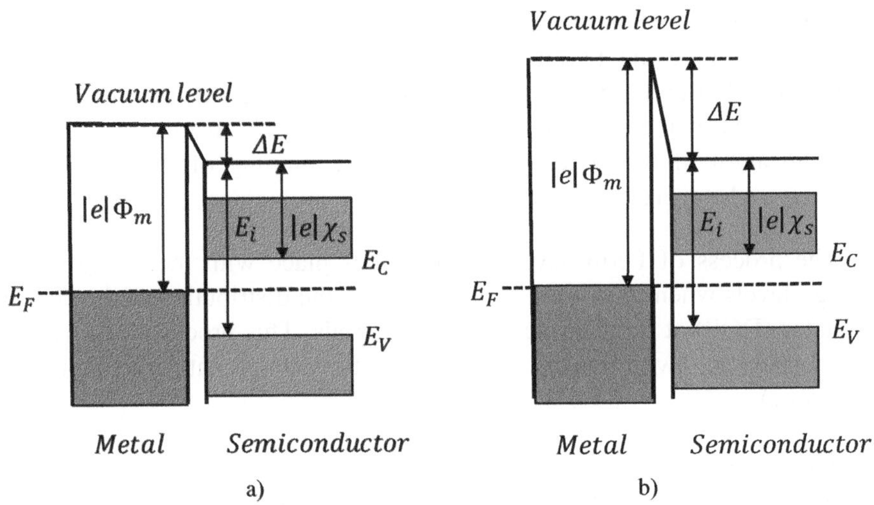

Figure 5.3. *Formation of the interface dipole layer: a) metal with a low work function; b) metal with a high work function. For a color version of this figure, see www.iste.co.uk/nguyen/electronics1.zip*

Several mechanisms may be responsible for the formation of the interface dipole (Ishii and Seki 2002): the transfer of charges at the interface, *the reduction of the electron cloud tail* at the metal surface by the organic adsorbate (known as the "push-back effect"), chemical interactions, the formation of surface states or the alignment of permanent dipoles of the organic material.

5.2.5. *Integer charge transfer process*

5.2.5.1. *Conductive substrate/organic semiconductor interface*

To explain Fermi-level pinning at the substrate/organic SC interface, a model called *integer charge transfer* (ICT), based on the transfer of electrons between different molecular levels, has been proposed (Fahlman *et al.* 2013). This model helps to explain the transition between two regimes of material contact, resulting either in the alignment of the vacuum-level or Fermi-level pinning. The substrate used is assumed to have a low interaction with the organic SC. In practice, it can be a conductive oxide (such as ITO and FTO) or an oxidized metal (such as Si/SiO_2 and Al/Al_2O_3) or any material whose surface is contaminated. Its work function is denoted by $|e|\Phi$ and its Fermi level by E_F.

Two energy states are defined for the charge transfer process. The first state, denoted as E_{ICT+}, is defined as the energy necessary to remove an electron from the molecule, which results in its complete relaxation (of charge and lattice deformation).

The second state, denoted as E_{ICT-}, is defined as the energy acquired when an electron is added to the molecule, which results in its complete relaxation (of charge and lattice deformation). These specifications apply to organic materials, since their structure is mechanically flexible and the addition or removal of charges causes deformations, such as the formation of polarons and bipolarons.

The charge transfer at a metal substrate/organic SC interface occurs once the work function of the metal is superior to the formation energy of charged states in the organic material. Even when surface states with low interaction are present, a charge transfer could occur by tunneling and create new energy states in the vicinity of the HOMO and LUMO bands. These new states become involved in the alignment process of the vacuum level instead of the HOMO and LUMO levels of the neutral molecule.

In this model, when the work function of the substrate varies between the two limits E_{ICT-} and E_{ICT+}, a transition occurs between two regimes: vacuum-level alignment and Fermi-level pinning, as shown in Figure 5.4.

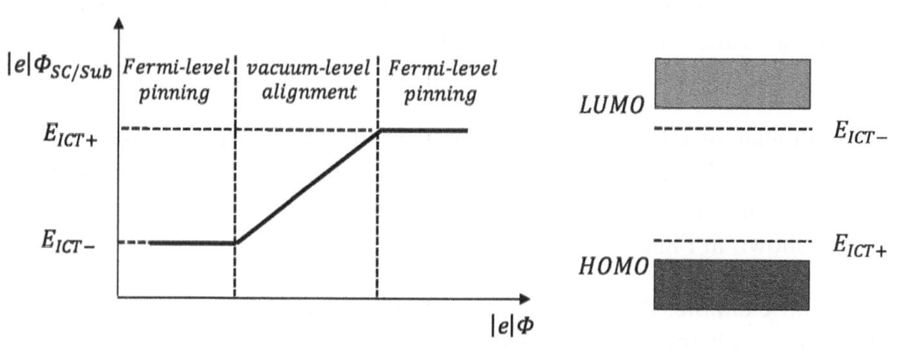

Figure 5.4. *Schematic representation of the ICT model with the transition from the vacuum-level alignment regime to the Fermi-level pinning regime. The levels EICT- and EICT+ are referenced at the vacuum level of the organic SC. For a color version of this figure, see www.iste.co.uk/nguyen/electronics1.zip*

The different behaviors of the contact between the substrate, assumed to be a metal, and the organic SC depend on the work function of the metal in the following way:

– for $|e|\Phi > E_{ICT+}$: the Fermi level E_F is pinned to the level E_{ICT+}. The potential barrier of the contact is independent of the nature of the metal layer. The mechanism of charge transfer when a contact is formed can be described as follows: initially, as $|e|\Phi > E_{ICT+}$, an electron transfer from the level E_{ICT+} to E_F will take place. The potential of the organic material gradually becomes positive, and the potential of the metal gradually becomes negative. This state creates an interface dipole that lowers the vacuum level of the organic material by an amount of energy ΔE. The process continues until the interface becomes balanced. At this point, the level E_{ICT+} aligns with the Fermi level E_F and $\Delta E = |e|\Phi - E_{ICT+}$;

– for $|e|\Phi < E_{ICT-}$: the Fermi level E_F is pinned to the level E_{ICT-}. The potential barrier of the contact is independent of the nature of the metal electrode. The mechanism for the charge transfer occurs in a way similar to that described in the previous case, with the inversion of the direction of the electron flow, that is, from the metal to the organic material. The potential of the metal gradually becomes positive and creates a positive interface dipole, which increases the vacuum level by an amount of energy ΔE. The process continues until the interface becomes balanced. At this point, the level E_{ICT-} aligns with the Fermi level E_F and $\Delta E = |e|\Phi - E_{ICT-}$;

– for $E_{ICT-} < |e|\Phi < E_{ICT+}$: the vacuum level is aligned. The potential barrier of the contact depends on the nature of the metal electrode. No transfer of charges occurs, and the equilibrium is achieved with the alignment of the vacuum level. Therefore, no interface dipoles are created. The slope of the line connecting the boundaries of the regimes is equal to 1.

A number of UPS experiments carried out with different substrates and different organic materials have demonstrated that this model is confirmed by the fact that it correlates with the measured values and the profile of the curves shown in Figure 5.4.

As an example, the polyfluorene APFO Green 1 deposited on different conductive substrates reveals two levels at 3.5 and 4.0 eV, indicating a transition from the vacuum-level alignment regime to the Fermi-level pinning regime (Crispin *et al.* 2006).

5.2.5.2. *Organic semiconductor/organic semiconductor interface*

In organic electronic devices, contacts between two organic SCs are frequently made. For example, the transport layers for electrons or holes are deposited onto or under the active layer to facilitate the transport of charge carriers.

When these layers come into contact, they form interfaces between organic SC 1 and organic SC 2. If no chemical reaction takes place between the layers, the contact between the organic materials is characterized by van der Waals bonding, that is, non-interactive bonds. Thus, the ICT charge transfer model can be applied to the organic/organic interface.

However, given that the transport layer is deposited on or under a conductive substrate (anode or cathode), the pinning levels of this layer, as well as that of the Fermi level of the substrate, will have an effect on the organic/organic contact. Therefore, it will be necessary to take into account the position of the energy levels of the organic SCs that are in contact and the position of the Fermi level of the conductive substrate. As an example, the study of the interface formed by the deposition of two films of 4,4'-N,N'-dicarbazoly biphenyl (CPB) and 4,4',4"-tris[3-methyl-phenyl(phenyl)amino]triphenylamine (m-MTDATA) on a substrate of poly(3,4-ethylene-dioxythiophene)-poly(perfluoroethylene sulfonic acid) (PEDOT-PFESA) previously deposited on an ITO substrate (Braun *et al.* 2007) has demonstrated that the alignment of energy levels at the interfaces of a multilayer depends on the relative position of the Fermi level of the conductive substrate and the molecular pinning level.

As a result, by changing the order in which the organic SCs are deposited (e.g. CBP/m-MTDATA instead of M-MTDATA/CBP), different alignments of energy levels can be obtained and therefore different electrical characteristics for the interfaces.

In conclusion, the integer charge transfer model allows most of the experimental results obtained for the metal/organic SC and organic SC/organic SC interfaces to be explained. It also provides useful information for selecting materials for making contact with the active layer of the organic device.

5.3. Surface characterization techniques

Before studying the characteristics of the interfaces between materials used in organic electronic devices, we will present the analytical techniques typically used to determine the physical parameters involved in the formation of the contact between two materials. For an SC, the gap value E_G can be measured by absorption spectroscopy by assimilating it to the optical gap. Other quantities of the energy band diagram such as the ionization energy E_i (IE or IP) of the SC and the work function of the metal $|e|\Phi_m$ can be obtained by *UV photoelectron spectroscopy*. The electron affinity $|e|\chi_S$ of the SC is determined by the difference between the energies E_i and E_G.

To study the changes in chemical bonds between atoms at the interface of the materials, *X-ray photoelectron spectroscopy* (XPS) is used. This technique also makes it possible to determine the chemical composition and the quantitative atomic concentration of the surface of the analyzed material. Optical microscopy is also useful for examining the morphology of the layer surface, which influences the formation of the contact between these layers. *Atomic force microscopy* (AFM) is used to measure the roughness of the surface at the nano-scale, and is particularly useful for examining contacts between materials. Other techniques for analyzing the surface can also be used, but they are less common than those we will examine here.

5.3.1. *Atomic force microscopy*

The technique of AFM allows us to determine the topology of a surface at the nano-scale. A microscope is equipped with a sharp tip mounted on the end of a cantilever. The apex or end point of the tip is very thin, which allows the examination of small size areas from a few nanometers to a few micrometers. The interaction between the apex and the surface atoms causes the cantilever to deform, which is sensed by the reflection of a laser beam pointed at the cantilever. The deviation of the laser beam allows the determination of the force of interaction and the topology of the surface.

There are two modes of AFM operation:

– *contact mode*, where the tip is brought very close to the surface of the sample. It enters into a repulsive interaction with the surface atoms, and the apex of the cantilever is dragged and scanned across the analyzed surface.

The movement of the apex allows the topology of the sample surface to be traced;

– *tapping mode*, where the tip oscillates at its resonant frequency with a fixed amplitude. When it undergoes an attractive interaction with the sample surface, the oscillation frequency is changed and the oscillation amplitude is reduced, resulting in a decrease in the amplitude of the movement of the cantilever. Using the same method as for the contact mode, changes in the amplitude of the cantilever are recorded and the topology of the surface to be studied is determined.

The analysis resolution depends on the size of the apex, which is subject to wear over time, especially when used in contact mode. The vertical resolution is of the order of nanometers and the lateral resolution is of the order of tens of nanometers.

5.3.2. *X-ray photoelectron spectroscopy*

X-ray photoelectron spectroscopy (XPS) is based on the photoelectric effect, which occurs when photons of an energy of $h\nu$ interact with the material and create photoelectrons, which will then be detected and analyzed. According to the principle of the conservation of energy, the kinetic energy E_k of electrons is expressed by:

$$E_k = h\,\nu - E_B - |e|\Phi \qquad [5.6]$$

where $|e|\Phi$ is the work function of the material, and E_B is the *binding energy* of the initial state of the electron with respect to the Fermi level. Photoelectrons are emitted from the energy levels occupied by electrons belonging to the VB of the SC, as well as by those belonging to the states at the core level region. They propagate to the surface of the sample and are then emitted into the vacuum, reaching a spectrometer that records the energy of the electrons. The XPS spectrum represents the distribution of the population of the photoelectrons as a function of their binding energy, or the density of states occupied by the material. Theoretically, the spectrum covers all the binding energies of the Fermi level, including the energy of the electrons in the valence region. For scattered electrons that do not have sufficient energy to overcome the work function $|e|\Phi$ of the material, they cannot escape the sample and will not be emitted. Therefore a phenomenon of *secondary electron cutoff* (SECO) occurs with a cutoff of the XPS spectrum; the measurement of the bonding energy E_{cutoff} corresponding to this cutoff allows the work function of the material to be determined. For a metal, the work function is:

$$|e|\Phi = |e|\Phi_m = h\nu - E_{cutoff} \quad [5.7]$$

Equation [5.7] is applicable to primary electrons, which are the electrons that are not elastically scattered after interacting with photons. Since the diffusion length of the electrons is very short (from 1 to 5 nm), the emitted photoelectrons are characteristic of the surface layers, and the measurements of the binding energy must be carried out under high vacuum ($< 10^{-9}$ mbar) so that the emitted electrons can reach the spectrometer. The metallic or organic films deposited by thermal evaporation are often obtained under much lower vacuum levels ($\sim 10^{-6}$ mbar), and the XPS analysis of these films generally reveals their surface to be slightly oxidized.

On the one hand, XPS spectra are used to determine the nature of the chemical bonds of the atom through a comparison of the spectrum of the element in a known chemical configuration with that of the same element in another configuration. The peak shift of an element, or the *chemical shift*, is a function of the nature of the compounds, and an analysis of the chemical shift can provide information that allows the molecular environments of the element in question to be determined. On the other hand, XPS spectra are also used for the semi-quantitative analysis of different atomic species found on the sample surface. The measurements of the peak intensity of these elements make it possible to determine the atomic concentrations, and consequently the chemical formula of the material surface.

5.3.3. *UV photoelectron spectroscopy*

The principle of UPS is the same as that of XPS, except for the fact that ultraviolet photons are used to excite electrons and enable their emission. The sources of the photons commonly used are the HEI (of energy $h\nu =$ 21.2 eV) and the HEII (of energy $h\nu = 40.8$ eV) from a helium discharge lamp. UPS spectra are limited to the energy domains of the valence region, but the accuracy of the measurement is better than that obtained by XPS spectra. The difference between the two techniques can be observed in the intensities of the spectra, explained by the effective photo-ionization cross-sections that vary with the energy of the photons used.

UPS spectra reflect the density of the occupied electronic states of the molecule and allow the steps of the interface formation to be analyzed and the process understood.

The different variables of the interfaces can be evaluated using the UPS spectrum of the material (or materials). In Figure 5.5a, the energy levels are shown for a metal with a work function $|e|\Phi_m$, with the metal Fermi level E_{Fm} used as the reference energy level. The value of the work function of the metal is determined by relationship [5.7], where the energy of the photons and the *cut-off energy* $E_{cutoffmetal}$ are known.

In Figure 5.5b, the energy levels for the metal and metal/organic SC interface are shown. The change ΔE in the work function is determined by the value of the cut-off energy $E_{cutoffmetal/SC}$. On the other hand, the potential barrier $|e|\Phi_{bp}$ for the injection of holes from the VB can be evaluated using the UPS spectrum. If ΔE and $|e|\Phi_{bp}$ are known, we can determine the ionization energy E_i.

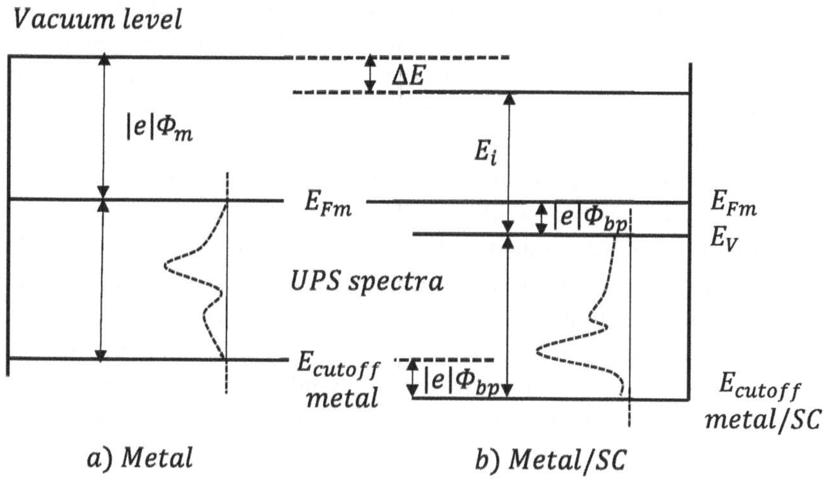

Figure 5.5. *Energy levels: a) in the metal; b) at the metal/SC interface*

5.4. Interface engineering

In an organic electronic device, the interface between the active layer and the other layers plays an important role in how the device operates, since the interface controls the injection and transport of the charges either directly (via direct contact) or indirectly (via contact through a buffer layer) with the electrodes. Thus, the interfaces must be carefully prepared to ensure that they behave reliably. In order to control the characteristics and quality of interfaces, certain techniques have been used to minimize the effects of chemical interactions between materials that modify the nature and structure of the interface layer. Interface engineering consists mainly of ensuring the stability of the contact structure between the different layers, in order to ensure the stability of the electronic processes in the device.

Here, we will examine two different approaches to adapting interfaces for the operation of devices.

5.4.1. *Inverted structure devices*

Consider the normal or direct structure of a basic organic solar cell. It consists of a transparent conductive electrode (such as ITO, which forms the anode) onto which a hole transport layer, the active layer, an electron transport layer and a metal layer (which forms the cathode) are then deposited. On the anode side, we know that the ITO is generally used for the electrode, and that the hole transport layer (HTL) must have band levels that are adapted to those of ITO to facilitate hole transport from the active layer to the anode. On the cathode side, taking into account the energy levels of the organic materials used for the electron transport layer (ETL), the metals chosen will have a low work function, such as calcium, barium or magnesium, to achieve contact between the cathode and the ETL. This choice is necessary if we wish to achieve an efficient electron transport from the cathode to the active layer. However, these materials are sensitive to oxygen and water vapor, and it will be necessary to encapsulate the device to slow down the oxidation of the cathode. To avoid the disadvantages of using an encapsulation, one simple solution is to use *the inverted structure* for the device, when this is possible.

In the inverted structure of an organic solar cell, the different layers are successively deposited on the anode (ITO) in the following order: the electron transport layer, the active layer, the hole transport layer and then the metal layer that forms the cathode. The main advantage of this structure over the normal structure is that it is possible to form the cathode by using metals with a high work function, such as gold or silver, which have a low susceptibility to oxidation. Therefore, these structures have a better stability than normal structures. In principle, devices with an inverted structure can operate in open air without needing to be protected by an encapsulation layer. Clearly, the materials used for the transport layers must be carefully selected to adapt to the energy characteristics of the electrodes and ensure the transport of holes (to the anode) and electrons (to the cathode), as shown in Figure 5.6b. Usually, the materials used for electron transport layers in the inverted structure are metal oxides, also known as *N-type oxides*, such as titanium dioxide (TiO_2), zinc oxide (ZnO) or cesium carbonate (Cs_2CO_3). For hole transport layers, also known as *P-type oxides*, transition metal oxides such as molybdenum trioxide (MoO_3), tungsten trioxide (WO_3) or vanadium pentoxide (V_2O_5) are used.

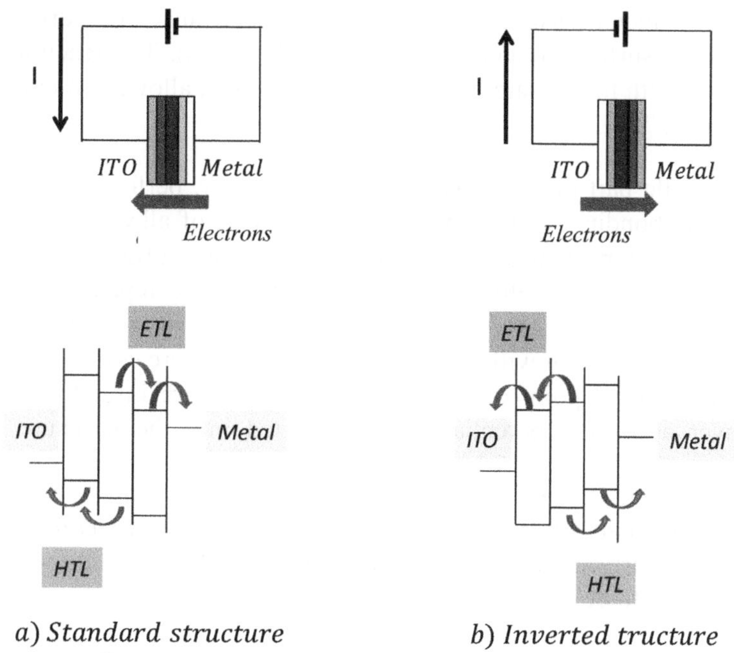

Figure 5.6. *Example of the structure of an organic solar cell: a) normal; b) inverted. For a color version of this figure, see www.iste.co.uk/nguyen/electronics1.zip*

Devices with inverted structures have roughly the same performance experimentally as those with a normal structure. On the contrary, the stability of these structures in ambient air is much better than that of those with a normal structure, thus highlighting the crucial role that the metal used for the electrodes plays in the degradation process of the devices.

5.4.2. *Self-assembled monolayers*

A *self-assembled monolayer* (SAM) is a layer of molecular thickness that forms spontaneously by adsorption without any external intervention. The assembly of molecules into films is possible thanks to the interaction forces between molecules. Each molecule of the layer consists of three parts, each of which performs a special function.

The lower part of the molecule, called the *head group*, performs the function of bonding with the substrate onto which the monolayer is deposited. The functional groups commonly used for this part are aromatic thiols (on a metal surface), phosphonic acids (on an oxide surface), and silanes (on a surface containing hydroxyl OH groups). The reaction of the head group with the substrate forms a covalent bond, allowing it to attach to the substrate.

The middle part of the molecule, called the *body*, performs the function of molecule bonding. Generally, the body consists of alkyl chains that are packed together by van der Waals forces. They are sufficient to form an ordered monolayer, but this layer is not sufficiently compact. To obtain a dense layer, it is necessary to use groups that develop strong bonds, such as hydrogen bonds, which will allow the film to have a more compact structure. From an electrical point of view, the alkyl chains are insulators, and the monolayer acts as a potential barrier inserted into the structure of the devices.

The upper part of the molecule, called the *tail* or *functional group*, performs the surface function of the monolayer. It allows the surface state of the monolayer to be controlled.

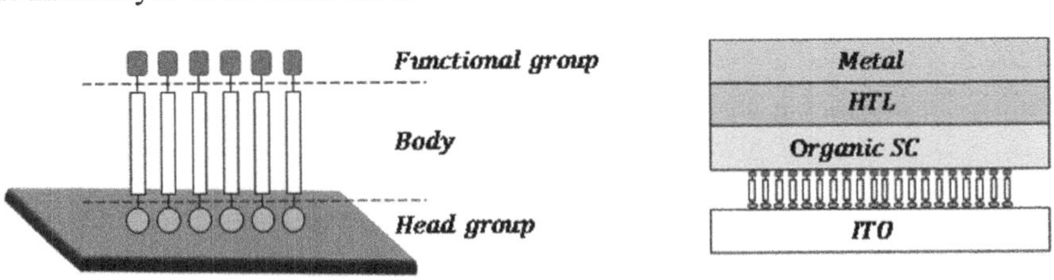

a) b)

Figure 5.7. *a) Structure of a self-assembled monolayer; b) example of the use of a self-assembled monolayer in an organic solar cell with an inverted structure. For a color version of this figure, see www.iste.co.uk/nguyen/electronics1.zip*

SAMs are generally insulators and can be used to modify the work function of contact metals and control the surface morphology of organic devices. When a SAM is inserted between the metal electrode and the organic film, an electrical dipole is formed at the interface, changing the distribution of charges on the surface of the metal. As a result, the work function of the contact metal will be changed. As we saw earlier, the value of the work function of the electrode can determine the process of the injection of charges in contact with the layers, leading either to the pinning of the Fermi level or the alignment of the vacuum level. The use of a SAM layer contributes to improving the injection of the charges and therefore the performance of the electronic device.

As an example, by modifying the work function of metal electrodes (gold and silver) using a SAM layer made up of alkanethiols, the current density of holes in a light-emitting diode using poly[2-methoxy-5-(2-ethyl-hexyloxy)-1,4-phenylene-vinyl] (MEH-PPV) as an emitter is increased by six orders of magnitude (De Boer *et al.* 2005). SAM layers are also used in organic transistors for controlling the charge transfer. The monolayer is deposited between an SC and an insulator, and forms an energy barrier between the two materials. Depending on the sign of the dipole induced by the SAM layer, the height of the barrier may increase or decrease, thus modifying the distribution of charge carriers in the SC (Fleischli *et al.* 2010).

The combination of the use of SAM with the inverted structure of the electronic device provides an effective tool for controlling interface processes, and allows the performance of the component to be improved. As an example, the study of inverted organic cells using an absorber of poly[4,8-bis-alkyloxybenzo[1,2-b:4,5-b']dithiophene-2,6-diyl-alt-4-(alkyl-1-one)thieno[3,4-b]thiophene-2,6-diyl] (PBDTTT-C)/[6,6]-phenyl-C61-butyric acid methyl ester (PCBM) has shown that the use of a monolayer of 3-aminopropyltriethoxysilane deposited between the ITO electrode and the active material improves the conversion efficiency by a factor of 8 (Song *et al.* 2013). This performance is achieved by reducing the work function of the ITO after the deposition of the monolayer. In addition, the stability of the cell when operating in ambient air has been significantly improved thanks to the SAM layer, which suppresses the diffusion of the metal atoms from the ITO electrode.

5.5. Conclusion

Interface processes in electronic devices are complex to analyze due to the fact that several physico-chemical phenomena can occur simultaneously during the formation of interfaces: chemical interactions, charge transfer and diffusion, and formation of clusters. The measurement techniques available for studying interfaces allow us to understand their structure and determine the conditions for applying the concept of vacuum-level alignment to the materials in contact. However, several questions remain to be clarified before we are able to reliably control the charge injection process, which is essential for the operation of the devices. These questions cover the electronic states of the surface, the recombination of excitons at the interface, the diffusion of excitons across the interface, etc.

Interface engineering allows the work function of the materials to be changed, and thus improve the charge injection process through the use of self-assembled monolayers. By using inverted structures, it is also possible to stabilize the structure and the operation of devices, allowing them to be used in ambient air and without encapsulation.

At the same time, all the parameters of these structures that allow the building of controlled interfaces, which are essential for the operation of the devices, have yet to be thoroughly understood and mastered.

Further theoretical and experimental investigations will be needed to better understand the interface processes of organic materials in order to improve the performances obtained in devices.

In Chapters 3, 4 and 5 of this first volume, we reviewed the fundamental processes of the operation of organic electronic devices. In Volume 2, we will study these devices in detail, with the goal of understanding their operation and determining the technological and economic parameters that allow them to be used and commercialized, responding to the needs and demands of the market.

List of Acronyms

General Terms

AFM	Atomic force microscopy
AMOLED	Active-matrix organic light-emitting diode
BHJ	Bulk heterojunction
CB	Conduction band
CELIV	Charge extraction by linearly increasing voltage
CHEMFET	Chemical field-effect transistor
CIE	*Commission internationale de l'éclairage* (International Commission on Illumination)
COP	Conference of the Parties
DLOS	Deep-level optical spectroscopy
DLTS	Deep-level transient spectroscopy
DOS	Density of states
EBL	Electron blocking layer
EPBT	Energy pay back time

EQE	External quantum efficiency
ETL	Electron transport layer
ETM	Electron transport material
FET	Field-effect transistor
GDM	Gaussian disorder model
GHG	Greenhouse gas
GWP	Global warming potential
HB	High barrier
HBL	Hole blocking layer
HI	Hysteresis index
HOMO	Highest occupied molecular orbital
HTL	Hole transport layer
HTM	Hole transport material
ICT	Integer charge transfer
IEA	International Energy Agency
IGFET	Insulated-gate field-effect transistor
IoT	Internet of Things
IPCE	Incident photon-to-current efficiency
LCD	Liquid crystal display
LCOE	Levelized cost of energy
LUMO	Lowest unoccupied molecular orbital
MOCVD	Metalorganic chemical vapor deposition

NIR	Near infrared
OE-A	Organic and Printed Electronics Association
OFET	Organic field-effect transistor
OLED	Organic light-emitting diode
OLET	Organic light-emitting transistor
OPV	Organic photovoltaic
OSC	Oxygen transmission rate
PCE	Power conversion efficiency
PMOLED	Passive-matrix organic light-emitting diode
Q-DLTS	Charge-based deep-level transient spectroscopy
RFID	Radio-frequency identification
SAM	Self-assembled monolayer
SC	Semiconductor
SCLC	Space-charge limited current
SCR	Space charge region
SED	Stretched exponential decay
SERS	Surface-enhanced Raman scattering
SVA	Solvent vapor annealing
TADF	Thermally activated delayed fluorescence
TCO	Transparent conducting oxides
TFL	Trap-filled limit
TFT	Thin-film transistor

TLC	Trapped-limited conduction
TOF	Time of flight
TOLED	Top-emitting organic light-emitting diode
TSC	Thermally stimulated current
TTA	Triplet–triplet annihilation
UHB	Ultra-high barrier
UPS	UV photoelectron spectroscopy
VB	Valence band
VOLET	Vertical organic light-emitting transistor
VRH	Variable-range hopping
WOLED	White organic light-emitting diode
WVTR	Water vapor transmission rate
XPS	X-ray photoelectron spectroscopy

Materials

Alq_3	tris(8-hydroxyquinoline)aluminum (III)
AZO	Aluminum-doped zinc oxide
BCB	benzocyclobutene
BCzVBi	4,4'-bis(9-ethyl-3-carbazovinylene'-1,1'-biphenyl
BDD	benzo[1,2-c:4,5-c']dithiophene-4,8-dione
BDPPV	benzodifurandione-based PPV
BDT	benzo[1,2-B:4,5-B']dithiophene

C-545T	10-(2-benzo-thiazolyl)-1,1,7,7-tetramethyl-2,3,6,7-tetrahydro-1H,5H,11H-[l]benzo-pyrano[6,7,8-ij]quinolizin-11
CFC-12	dichlorofluoromethane
CN-PPVp	poly(cyano-phenylene vinylene)
CPB	4,4'-N,N'-dicarbazolyl biphenyl
CYEP	cyanoethylpullulane
CYTOP	poly(perfluoroethylene-co-butenyl vinyl ether)
DCJTB	4-(dicyanomethylene)-2-t-butyl-6-(1,1,7,7-tetramethyljulolidyl-9-enyl)-4H-pyran
DCJTI	4-(dicyanomethylene)-2-i-propyl-6-(1,1,7,7-tetramethyljulolidyl-9-enyl)-4H-pyran
DH-4T	α,ω-dihexyl-quaterthiophene
DH-6T	dihexyl-sexithiophene
DHTBT	4,7-di(4-hexylthien-2-yl)-benzothiadiazole
DM	N2,N2',N7,N7'-tetrakis (9,9-dimethyl-9H-fluorene-2-yl)- N2,N2'N7,N7'-tetrakis (4-methoxy phenyl)-9,9'-spirobi[fluorene]-2,2',7,7'-tetraamine
DMF	N,n-dimethylformamide
DMSO	dimethylsulfoxide
DMZ	dimethylzinc
DPVBi	4,4-bis(2,2-diphenylvinyl)biphenyl
F8T2	poly[9,9-dioctylfluorene-co-bithiophene]
FIrpic	bis[2-(4,6-difluorophenyl)pyridinato-C2,N](picolinato)iridium (III)

FTO	Fluorine-doped tin oxide
GO	Graphene oxide
HMDS	hexamethyldisilazane
IPA	isopropanol
Ir(HFP)	tris[2,5-bis-2'(9',9'-dihexylfluorene)pyridine] iridium (III)
ITO	Indium tin oxide
m-MTDATA	4,4',4"-tris[3-methyl-phenyl(phenyl)amino] triphenylamine
MAPI	methylammonium lead iodide perovskite
MDMO-PPV	poly(2-methoxy-5-(3',7'-dimethyl-octyloxy)-1,4-phenylene vinylene)
MEH-PPV	poly[2-methoxy-5-(2-ethyl-hexyloxy)-1,4-phenylene-vinyl]
MWCNT	Multi-walled carbon nanotubes
NDI	naphthalene diimide
OTS	octadecyltrichlorosilane
P3HT	poly(hexylthiophene)
PANI	polyaniline
PAT	poly(alkylthiophene)
PBDB-T	poly[(2,6-(4,8-bis(5-(2-ethylhexyl)thiophen-2-yl)-benzo[1,2-b:4,5-b']dithiophene))-alt-(5,5-(1',3'-di-2-thienyl-5',7'-bis(2-ethylhexyl)benzo[1',2'-c:4',5'-c']dithiophene-4,8-dione))]

PBDTTT-C	poly[4,8-bis-alkyloxybenzo[1,2-b:4,5-b']dithiophene-2,6-diyl-alt-4-(alkyl-1-one)thieno[3,4-b]thiophene-2,6-diyl]
PBTTT	poly(2,5-bis(3-alkylthiophen-2-yl)thieno(3,2-b)thiophene)
Pc	phthalocyanine
PCBM	[6,6]-phenyl-C61-butyric acid methyl ester
PCDTBT	poly[N-9'-heptadecanyl-2,7-carbazole-alt-5,5-(4,7-di-2-thienyl-2',1',3'-benzo-thiadiazole)]
PCPDTBT	poly[2,6-(4,4-bis-(2-ethylhexyl)-4H-cyclopenta[2,1-b;3,4-b']dithiophene)-alt-4,7-(2,1,3-benzothiadiazole)]
PDI	perylene diimide
PDMS	poly(dimethysiloxane)
PEDOT:PSS	poly(3,4-ethylenedioxythiophene)-poly(styrenesulfonate)
PEDOT-PFESA	poly(3,4-ethylene-dioxythiophene)-poly(perfluoroethylene sulfonic acid)
PET	poly(ethylene terephthalate)
PFH-MEH	poly[(9,9-dihexyloxyfluoren-2,7-diyl)-alt-co-(2-methoxy-5-{2-ethylhexyloxy}phenylen-1,4-diyl)]
PFO	polyfluorene
PGMEA	propylene glycol-monomethyl ether acetate
PI	polyimide
PIF	poly(3,9-di-*t*-butylindeno[1,2-b]fluorene)
PITN	poly(isathianapthene)

PMMA	poly(methyl methacrylate)
POSS	Polyhedral oligomeric silsesquioxane
PPP	poly(p-phenylene)
PPV	poly(phenylene vinylene)
PPy	polypyrrole
PSBF	poly(spirobifluorene)
PTAA	poly(triaryl amine)
PTB7-TH	poly[4,8-bis(5-(2-ethylhexyl)thiophen-2-yl)benzo[1,2-b;4,5-b']dithiophene-2,6-diyl-alt-(4-(2-ethylhexyl)-3-fluorothieno[3,4-b]thiophene-)-(2-carboxylate-2-6-diyl)]
PTQ	poly[[6,7-difluoro[(2-hexyldecyl)oxy]-5,8-quinoxalinediyl]-2,5-thiophenediyl]
PVA	poly(vinyl alcohol)
PVK	poly(n-vinylcarbazole)
PVP	poly-4-vinylphenol
SO	dibenzothiophene-*S,S*-dioxide
spiro-OMeTAD	2,2',7,7'-tetrakis-(N,N-di-4-methoxyphenylamino)-9,9'-spirobifluorene
TBADN	2-tert-butyl-9,10-di(naphth-2-yl)anthracene
Tips-pentacene	6,13-bis (triisopropyl-silylethynyl)pentacene
TPA	triphenylamine

References

Alvi, F., Ram, M.K., Gomez, H., Joshi, R.K., and Kumar, A. (2010). Evaluating the chemio-physio properties of novel zinc oxide-polyaniline nanocomposite polymer films. *Polymer Journal*, 42(12), 935–940.

Bardeen, J. (1947). Surface states and rectification at a metal semi-conductor contact. *Physical Review*, 71(10), 717–727.

Bässler, H. (1993). Charge transport in disordered organic photoconductors: A Monte Carlo simulation study. *Physica Status Solidi (B)*, 175(1), 15–56.

Bässler, H. (2000). Charge transport in random organic semiconductors. In *Semiconducting Polymers*, Hadziioannou, G. and van Hutten, P.F. (eds). Wiley-VCH, Weinheim.

Bisquert, J., Garcia-Belmonte, G., Munar, A., Sessolo, M., Soriano, A., and Bolink, H.J. (2008). Band unpinning and photovoltaic model for P3HT:PCBM organic bulk heterojunctions under illumination. *Chemical Physics Letters*, 465(1–3), 57–62.

Blades, C.D.J. and Walker, A.B. (2000). Simulating organic light-emitting diodes. *Synthetic Metals*, 111–112, 335–340.

Braun, S., De Jong, M.P., Osikowicz, W., and Salaneck, W.R. (2007). Influence of the electrode work function on the energy level alignment at organic-organic interfaces. *Applied Physics Letters*, 91, 202108.

Braun, S., Salaneck, W.R., and Fahlman, M. (2009). Energy-level alignment at organic/metal and organic/organic interfaces. *Advanced Materials*, 21(14–15), 1450–1472.

Brazowskii, S. and Kirova, N.N. (1993). Towards the theory of metal-polymer contact. *Synthetic Metals*, 57, 4385–4392.

Brédas, J.L., Chance, R.R., Silbey, R., Nicolas, G., and Durand, P. (1982). Valence effective Hamiltonian study of the electronic structure of poly(p-phenylene) and poly(p-phenylene sulfide). *The Journal of Chemical Physics*, 77(1), 371–378.

Burrows, P.E., Bulovic, V., Forrest, S.R., Sapochak, L.S., McCarty, D.M., and Thompson, M.E. (1994). Reliability and degradation of organic light emitting devices. *Applied Physics Letters*, 65(23), 2922–2924.

Burtone, L., Ray, D., Leo, K., and Riede, M. (2012). Impedance model of trap states for characterization of organic semiconductor devices. *Journal of Applied Physics*, 111, 604503.

Campbell, A.J., Bradley, D.D.C., Werner, E., and Brütting, W. (2000). Transient capacitance measurements of the transport and trap states distributions in a conjugated polymer. *Organic Electronics: Physics, Materials, Applications*, 1(1), 21–26.

Cowell, T.A.T. and Woods, J. (1967). The evaluation of thermally stimulated current curves. *British Journal of Applied Physics*, 18, 1045–1051.

Crispin, A., Crispin, X., Fahlman, M., Berggren, M., and Salaneck, W.R. (2006). Transition between energy level alignment regimes at a low band gap polymer-electrode interfaces. *Applied Physics Letters*, 89, 213503.

Curioni, A. and Andreoni, W. (2000). Organic-cathode interface in Alq3-based organic light-emitting devices: New insights from abinitio molecular dynamics. *Synthetic Metals*, 111, 299–301.

De Boer, B., Hadipour, A., Mandoc, M.M., and Blom, P.W.M. (2005). Tuning of metal work function with self-assembled monolayers. *Materials Research Society Symposium Proceedings*, 871(5), 189–196.

De Jong, M.P., Van Ijzendoorn, L.J., and De Voigt, M.J.A. (2000). Stability of the interface between indium-tin-oxide and poly(3,4-ethylenedioxythiophene)/poly(styrenesulfonate) in polymer light-emitting diodes. *Applied Physics Letters*, 77(14), 2255–2257.

Dollinger, F., Nehm, F., Müller-Meskamp, L., and Leo, K. (2017). Laminated aluminum thin-films as low-cost opaque moisture ultra-barriers for flexible organic electronic devices. *Organic Electronics: Physics, Materials, Applications*, 46, 242–246.

Fahlman, M., Sehati, P., Osikowicz, W., Braun, S., De Jong, M.P., and Brocks, G. (2013). Photo-electron spectroscopy and modeling of interface properties related to organic photovoltaic cells. *Journal of Electron Spectroscopy and Related Phenomena*, 190, 33–41.

Fleischli, F.D., Suárez, S., Schaer, M., and Zuppiroli, L. (2010). Organic thin-film transistors: The passivation of the dielectric-pentacene interface by dipolar self-assembled monolayers. *Langmuir*, 26(18), 15044–15049.

Gaal, M., List, E.J.W., and Scherf, U. (2003). Excimers or emissive on-chain defects? *Macromolecules*, 36(11), 4236–4237.

Gao, Y. (2010). Surface analytical studies of interfaces in organic semiconductor devices. *Materials Science and Engineering R: Reports*, 68(3), 39–87.

Heeger, A.J., Kivelson, S., Schrieffer, J.R., and Su, W.P. (1988). Solitons in conducting polymers. *Reviews of Modern Physics*, 60(3), 781–850.

Hu, T., Zhang, F., Xu, Z., Zhao, S., Yue, X., and Yuan, G. (2009). Effect of UV-ozone treatment on ITO and post-annealing on the performance of organic solar cells. *Synthetic Metals*, 159(7–8), 754–756.

Hwang, W. and Kao, K.C. (1976). Studies of the theory of single and double injections in solids with a Gaussian trap distribution. *Solid State Electronics*, 19(12), 1045–1047.

Ip, J., Nguyen, T.P., and Le Rendu, P. (2003). An X-ray photoelectron spectroscopy study of the interface formed between ITO and 4,4′-bis(4-dimethyl amino styryl)benzene based light-emitting diode. *Synthetic Metals*, 138(1–2), 107–111.

Ishii, B.H. and Seki, K. (2002). Interface materials for organic solar cells. In *Conjugated Polymer and Molecular Interfaces*, Salaneck, W.R., Seki, K., Kahn, A., and Pireaux, J.-J. (eds). Marcel Dekker, New York, 293–349.

Ishii, B.H., Sugiyama, K., Ito, E., and Seki, K. (1999). Energy level alignment and interfacial electronic structures at organic/metal and organic/organic interfaces. *Advanced Materials*, 11(8), 605–625.

Juska, G., Arlauskas, K., and Viliūnas, M. (2000). Extraction current transients: New method of study of charge transport in microcrystalline silicon. *Physical Review Letters*, 84(21), 4946–4949.

Knapp, E., Häusermann, R., Schwarzenbach, H.U., and Ruhstaller, B. (2010). Numerical simulation of charge transport in disordered organic semiconductor devices. *Journal of Applied Physics*, 108, 054504.

Knupfer, M. (2003). Exciton binding energies in organic semiconductors. *Applied Physics A: Materials Science and Processing*, 77(5), 623–626.

Kobayashi, T., Bando, M., Kimura, N., Shimizu, K., Kadono, K., Umezu, N., Miyahara, K., Hayazaki, S., Nagai, S., Mizuguchi, Y., Murakami, Y., and Hobara, D. (2013). Production of a 100-m-long high-quality graphene transparent conductive film by roll-to-roll chemical vapor deposition and transfer process. *Applied Physics Letters*, 102(2), 023112.

Leng, J.M., Jeglinski, S., Wei, X., Benner, R.E., Vardeny, Z.V., Guo, F., and Mazumdar, S. (1994). Optical probes of excited states in poly(p-phenylenevinylene). *Physical Review Letters*, 72(1), 156–159.

Malard, L.M., Pimenta, M.A., Dresselhaus, G., and Dresselhaus, M.S. (2009). Raman spectros-copy in graphene. *Physics Reports*, 473(5–6), 51–87.

Marks, R.N., Halls, J.J.M., Bradley, D.D.C., Friend, R.H., and Holmes, A.B. (1994). The photo-voltaic response in poly(p-phenylene vinylene) thin-film devices. *Journal of Physics: Condensed Matter*, 6(7), 1379–1394.

Miller, A. and Abrahams, E. (1960). Impurity conduction at low concentrations. *Physical Review*, 120(3), 745–755.

Milliron, D.J., Hill, I.G., Shen, C., Kahn, A., and Schwartz, J. (2000). Surface oxidation activates indium tin oxide for hole injection. *Journal of Applied Physics*, 87(1), 572–576.

Mönch, W. (1990). On the physics of metal-semiconductor interfaces. *Reports on Progress in Physics*, 53(3), 221–278.

Mott, N. (1993). *Conduction in Non-Crystalline Materials*. Clarendon Press, Oxford.

Nakano, Y., Noda, K., Fujikawa, H., Morikawa, T., Ohwaki, T., and Taga, Y. (2006). Trap levels in tris(8-hydroxyquinoline) aluminum studied by deep-level optical spectroscopy. *Applied Physics Letters*, 88, 252104.

Neugebauer, S., Rauh, J., Deibel, C., and Dyakonov, V. (2012). Investigation of electronic trap states in organic photovoltaic materials by current-based deep level transient spectroscopy. *Applied Physics Letters*, 100, 263304.

Nguyen, T.P., and Ip, J. (2002). Study of indium tin oxide -4,4′-bis(4-dimethylaminostyryl) benzene interface: An X ray photoelectron spectroscopy investigation. *Applied Surface Science*, 195(1–4), 251–262.

Nguyen, T.P. and Renaud, C. (2009). Interface effects on the defect state formation in organic devices. *Organic Photonic Materials and Devices XI*, 721314.

Nguyen, T.P., Tran, V.H., Destruel, P., and Oelkrug, D. (1999). Optical spectroscopic investigations of phenylene vinylene oligomers. *Synthetic Metals*, 101(1), 633–634.

Nguyen, T.P., Le Rendu, P., Dinh, N.N., Fourmigué, M., and Mézière, C. (2003). Thermal and chemical treatment of ITO substrates for improvement of OLED performance. *Synthetic Metals*, 138(1–2), 229–232.

Nguyen, T.P., Ip, J., Gaudin, O., and Jackman, R.B. (2004). Determination of localized trap parameters in organic semiconductors using charge based deep level transient spectroscopy (Q-DLTS). *The European Physical Journal – Applied Physics*, 27, 201–222.

Nguyen, T.P., Renaud, C., Reisdorffer, F., and Wang, L. (2012). Degradation of phenyl C61 butyric acid methyl ester: Poly (3-hexylthiophene) organic photovoltaic cells and structure changes as determined by defect investigations. *Journal of Photonics for Energy*, 2(1), 021013.

Nicolai, H.T., Mandoc, M.M., and Blom, P.W.M. (2011). Electron traps in semiconducting polymers: Exponential versus Gaussian trap distribution. *Physical Review B – Condensed Matter and Materials Physics*, 83(19), 1–5.

Nunes De Carvalho, C., Botelho Do Rego, A.M., Amaral, A., Brogueira, P., and Lavareda, G. (2000). Effect of substrate temperature on the surface structure, composition and morphology of indium-tin oxide films. *Surface and Coatings Technology*, 124(1), 70–75.

Ostroverkhova, O. (2016). Organic optoelectronic materials: Mechanisms and applications. *Chemical Reviews*, 116(22), 13279–13412.

Park, J.S., Chae, H., Chung, H.K., and Lee, S.I. (2011). Thin film encapsulation for flexible AM-OLED: A review. *Semiconductor Science and Technology*, 26(3), 034601.

Parker, I.D. (1994). Carrier tunneling and device characteristics in polymer light-emitting diodes. *Journal of Applied Physics*, 75(3), 1656–1666.

Renaud, Ć. and Nguyen, T.P. (2008). Study of trap states in polyspirobifluorene based devices: Influence of chromophore addition. *Journal of Applied Physics*, 104, 113705.

Renaud, Ć. and Nguyen, T.P. (2009). Study of trap states in polyspirobifluorene based devices: Influence of aging by electrical stress. *Journal of Applied Physics*, 106, 053707.

Renaud, Ć. and Nguyen, T.P. (2010). Identification of the nature of trapping centers in polys-pirobifluorene based diodes by using electrical characterization. *Journal of Applied Physics*, 107, 124505.

Schafferhans, J., Baumann, A., Wagenpfahl, A., Deibel, C., and Dyakonov, V. (2010). Oxygen doping of P3HT:PCBM blends: Influence on trap states, charge carrier mobility and solar cell performance. *Organic Electronics: Physics, Materials, Applications*, 11(10), 1693–1700.

Scher, H. and Montroll, E.W. (1975). Anomalous transit-time dispersion in amorphous solids. *Physical Review B*, 12(6), 2455–2477.

Singh, D.K., Iyer, P.K., and Giri, P.K. (2012). Role of molecular interactions and structural defects in the efficient fluorescence quenching by carbon nanotubes. *Carbon*, 50(12), 4495–4505.

Song, M., Kang, J.W., Kim, D.H., Kwon, J.D., Park, S.G., Nam, S., Jo, S., Ryu, S.Y., and Kim, C.S. (2013). Self-assembled monolayer as an interfacial modification material for highly efficient and air-stable inverted organic solar cells. *Applied Physics Letters*, 102, 143303.

Spano, F.C. and Silva, C. (2014). H- and J-Aggregate behavior in polymeric semiconductors. *Annual Review of Physical Chemistry*, 65(1), 477–500.

Spano, F.C., Clark, J., Silva, C., and Friend, R.H. (2009). Determining exciton coherence from the photoluminescence spectral line shape in poly(3-hexylthiophene) thin films. *Journal of Chemical Physics*, 130(7), 1–16.

Vinoth, R., Babu, S.G., Bharti, V., Gupta, V., Navaneethan, M., Bhat, S.V., Muthamizhchelvan, C., Ramamurthy, P.C., Sharma, C., Aswal, D.K., Hayakawa, Y., and Neppolian, B. (2017). Ruthenium based metallopolymer grafted reduced graphene oxide as a new hybrid solar light harvester in polymer solar cells. *Scientific Reports*, 7(43133), 1–14.

Wang, H.Y., Wang, F.M., Wang, Y.Y., Wan, C.C., Hwang, B.J., Santhanam, R., and Rick, J. (2011). Electrochemical formation of Pt nanoparticles on multiwalled carbon nanotubes: Useful for fabricating electrodes for use in dye-sensitized solar cells. *Journal of Physical Chemistry C*, 115(16), 8439–8446.

Weaver, M.S., Michalski, L.A., Rajan, K., Rothman, M.A., Silvernail, J.A., Brown, J.J., Burrows, P.E., Graff, G.L., Gross, M.E., Martin, P.M., Hall, M., Mast, E., Bonham, C., Bennett, W., and Zumhoff, M. (2002). Organic light-emitting devices with extended operating lifetimes on plastic substrates. *Applied Physics Letters*, 81(16), 2929–2931.

Yan, M., Rothberg, L.J., Papadimitrakopoulos, F., Galvin, M.E., and Miller, T.M. (1994). Defect quenching of conjugated polymer luminescence. *Physical Review Letters*, 73(5), 744–747.

Yang, J. and Shen, J. (1999). Effects of discrete trap levels on organic light-emitting diodes. *Journal of Applied Physics*, 85(5), 2699–2705.

Yi, G., Wang, C., and Park, W.I. (2005). ZnO nanorods: Synthesis, characterization and applications. *Semiconductor Science and Technology*, 20, S22–S34.

Zhang, D., Ryu, K., Liu, X., Polikarpov, E., Tompson, M., and Zhou, C. (2006). Transparent, conductive and flexible carbon nanotube films and their application in organic light-emitting diodes. *Materials Research Society Symposium Proceedings*, 936, 19–24.

Zou, J.P., Le Rendu, P., Musa, I., Yang, S.H., Dan, Y., That, C.T., and Nguyen, T.P. (2011). Investigation of the optical properties of polyfluorene/ZnO nanocomposites. *Thin Solid Films*, 519(12), 3997–4003.

Index

G, H, I

L, M, P

R, S, T

U, V, W, X

Other titles from

ISTE

in

Electronics Engineering

2020

CAPPY Alain
Neuro-inspired Information Processing

FOUQUET François
Noise in Radio-Frequency Electronics and its Measurement

MAHDOUM Ali
CAD of Circuits and Integrated Systems

2018

MURET Pierre
Fundamentals of Electronics 2: Continuous-time Signals and Systems
Fundamentals of Electronics 3: Discrete-time Signals and Systems, and Quantized Level Systems

2017

MURET Pierre
Fundamentals of Electronics 1: Electronic Components and Elementary Functions

BUCCI Davide
Analog Electronics for Measuring Systems

2016

BAUDRAND Henri, TITAOUINE Mohammed, RAVEU Nathalie
The Wave Concept in Electromagnetism and Circuits: Theory and Applications

FANET Hervé
Ultra Low Power Electronics and Adiabatic Solutions

NDJOUNTCHE Tertulien
Digital Electronics 1: Combinational Logic Circuits
Digital Electronics 2: Sequential and Arithmetic Logic Circuits
Digital Electronics 3: Finite-state Machines

2015

DURAFFOURG Laurent, ARCAMONE Julien
Nanoelectromechanical Systems

2014

APPRIOU Alain
Uncertainty Theories and Multisensor Data Fusion

CONSONNI Vincent, FEUILLET Guy
Wide Band Gap Semiconductor Nanowires 1: Low-Dimensionality Effects and Growth
Wide Band Gap Semiconductor Nanowires 2: Heterostructures and Optoelectronic Devices

GAUTIER Jean-Luc
Design of Microwave Active Devices

LACAZE Pierre Camille, LACROIX Jean-Christophe
Non-volatile Memories

TEMPLIER François
OLED Microdisplays: Technology and Applications

THOMAS Jean-Hugh, YAAKOUBI Nourdin
New Sensors and Processing Chain

2013

COSTA François, GAUTIER Cyrille, LABOURE Eric, REVOL Bertrand
Electromagnetic Compatibility in Power Electronics

KORDON Fabrice, HUGUES Jérôme, CANALS Agusti, DOHET Alain
Embedded Systems: Analysis and Modeling with SysML, UML and AADL

LE TIEC Yannick
Chemistry in Microelectronics

2012

BECHERRAWY Tamer
Electromagnetism: Maxwell Equations, Wave Propagation and Emission

LALAUZE René
Chemical Sensors and Biosensors

LE MENN Marc
Instrumentation and Metrology in Oceanography

SAGUET Pierre
Numerical Analysis in Electromagnetics: The TLM Method

2011

ALGANI Catherine, RUMELHARD Christian, BILLABERT Anne-Laure
Microwaves Photonic Links: Components and Circuits

BAUDRANT Annie
Silicon Technologies: Ion Implantation and Thermal Treatment

DEFAY Emmanuel
Integration of Ferroelectric and Piezoelectric Thin Films: Concepts and Applications for Microsystems

DEFAY Emmanuel
Ferroelectric Dielectrics Integrated on Silicon

BESNIER Philippe, DÉMOULIN Bernard
Electromagnetic Reverberation Chambers

LANDIS Stefan
Nano-lithography

2010

LANDIS Stefan
Lithography

PIETTE Bernard
VHF / UHF Filters and Multicouplers

2009

DE SALVO Barbara
Silicon Non-volatile Memories / Paths of Innovation

DECOSTER Didier, HARARI Joseph
Optoelectronic Sensors

FABRY Pierre, FOULETIER Jacques
Chemical and Biological Microsensors / Applications in Fluid Media

GAUTIER Jacques
Physics and Operation of Silicon Devices in Integrated Circuits

MOLITON André
Solid-State Physics for Electronics

PERRET Robert
Power Electronics Semiconductor Devices

SAGUET Pierre
Passive RF Integrated Circuits

2008

CHARRUAU Stéphane
Electromagnetism and Interconnections

2007

RIPKA Pavel, TIPEK Alois
Modern Sensors Handbook

Printed and bound by CPI Group (UK) Ltd, Croydon, CR0 4YY

15/09/2026

14962349-0001